SPACE SAILING

Square-Rigged Sailing Ship in Earth Orbit

SPACE SAILING

Jerome L. Wright
General Astronautics Corporation, Titusville, Florida, USA

GORDON AND BREACH SCIENCE PUBLISHERS
Switzerland • Australia • Belgium • France • Germany • Great Britain
India • Japan • Malaysia • Netherlands • Russia • Singapore • USA

First published 1992
Second printing 1993

Gordon and Breach Science Publishers

Y. Parc
Chemin de la Sallaz
1400 Yverdon, Switzerland

Post Office Box 90
Reading, Berkshire RG1 8JL
Great Britain

Private Bag 8
Camberwell, Victoria 3124
Australia

3-14-9, Okubo
Shinjuku-ku, Tokyo 169
Japan

58, rue Lhomond
75005 Paris
France

Emmaplein 5
1075 AW Amsterdam
Netherlands

Glinkastrasse 13-15
0-1086 Berlin
Germany

820 Town Center Drive
Langhorne, Pennsylvania 19047
United States of America

Library of Congress Cataloging-in-Publication Data

Wright, Jerome, 1944-
Space sailing / Jerome Wright.
p. cm.
Includes index.
ISBN 2-88124-803-9 (hardcover)
ISBN 2-88124-842-X (softcover)
1. Solar sails. 2. Space flight. I. Title.
TL783.9.W75 1992
629.47'54--dc20 91-38093
CIP

In memory of
William F. Carroll
who knew why we are leaving Earth

CONTENTS

PREFACE

The description of electromagnetic fields and radiation was developed by James Clerk Maxwell. His theory predicted that a pressure results from the radiation. When it was discovered that light is actually electromagnetic radiation, it was deduced that light must exert pressure. From this, Svant Arrhenius predicted the possibility of light pressure driving life spores across space, which was the first suggestion that light could transport things through space.

The concept of using light to propel ships, employing photonic engines, was proposed by Konstantine Tsiolkovsky. The first known scheme to use sunlight to drive ships – solar sailing – was invented by Frederik Tsander.

Published articles on solar sailing began appearing in the United States in the 1950s by authors unaware of the earlier work by Tsander. Carl Wiley, writing as Russell Saunders, and Richard Garwin were the first American authors.

As interest in the concept was increasing, I began a study of solar sailing while at Battelle Memorial Institute in Columbus, Ohio. At the suggestion of Donald Edgecombe, I developed a trajectory that showed a possibility for a solar sail spacecraft to rendezvous with Halley's comet during its 1986 passage.

Several people at the Jet Propulsion Laboratory in Pasadena, California, became interested in the proposal for a rendezvous with Halley's comet. In particular, Louis Friedman and Chauncey Uphoff became strong supporters, which led to my moving to JPL to lead the mission design on solar sails.

Bruce Murray, the director of JPL, was very enthusiastic about the solar sail and the Halley Rendezvous mission, as we came to call it. The effort at JPL during the following year did far more to develop solar sail technology than any other effort to date. Several solar sail concepts underwent both inventive processes and critical examination. At the end of the effort, funding from the National Aeronautics and Space Administration was abruptly cut. A modest effort was made to complete the final reports, but very few of the technical results have ever been published publicly.

At the end of a competitive effort in 1977, a solar electric propulsion concept was selected instead of the solar sail for the Halley Rendezvous mission, primarily because of technology maturity. Not long after, the Halley Rendezvous mission was dropped by NASA. Eventually, even the solar electric propulsion development was dropped as planetary exploration activity declined even further in the face of space shuttle and space station activities. The history and politics of the solar sail effort are described by Friedman in his book *Star Sailing*, published by John Wiley & Sons, New York.

For a while, the only solar sail research was at the University of Utah, where Professor Gary Flandro and a group of engineering students attempted to develop a sail deployment experiment to go on a space shuttle flight. In 1979, the World Space Foundation began a solar sail experiment that resulted in a small sail being constructed and deployed in a successful ground test. The foundation is presently attempting to reach an agreement with NASA and other launch organizations to launch two larger sails, one as a deployment test in space, and the other as an operational spacecraft. An international effort is in progress to launch a three-sail spacecraft, but its future is very uncertain.

The technical work done to date shows that sailing ships are practical, low-cost concepts. Sails appear to have definite potential as useful devices for space transportation. The work done at JPL and other institutions and companies was an important step towards the development of that potential.

The purpose of this book is to make available to the public some of the technical results of the design effort conducted at JPL. References given in this book are sparse, generally because much of the source material is in the form of internal reports and memorandums that have never been published.

ACKNOWLEDGEMENTS

The design effort at JPL in 1976 and 1977 was successful in developing a technology base for sailing spacecraft because of the dedicated efforts of many people. A list of all contributors is far too large to include here, but Louis Friedman, William Layman, William Carroll, John Kevit, James Stevens, Carl Sauer, Richard MacNeal and John Hedgepath should be mentioned.

Robert Forward generously provided material for this book and reviewed parts of it.

The following companies made direct contributions to the solar sail effort at JPL:

Astro Research	General Research
Battelle Columbus	ILC-Dover
Boeing	Kinetic Coatings
Boeing-Vertol	MacNeal Schwendler
Ciba-Geigy	Martin Marietta
Draper Laboratory	Rockwell
Du Pont	Schweitzer
Endurex	Sheldahl
General Dynamics	TRW

1. INTRODUCTION TO SAILING

Stars are immense thermal engines driven by nuclear reactions. Their released energy leaves primarily in the form of electromagnetic radiation: light. Planets intercept only an extremely small fraction of this energy, while the rest is lost to cosmic space.

Spacecraft using large sails can use some of this otherwise lost energy. Solar sails use sunlight in a direct manner to move through space, while laser sails use solar energy which has been converted into a laser beam. Two characteristic of such spacecraft are the use of large, passive reflectors and the complete absence of onboard generated power to obtain a propulsive effect.

Solar sails are large, wispy films drifting through space like apparitions, without noise and vibration of engines. They are able to move themselves through solar systems for decades as cargo carriers, space liners, and explorers.

What if we build such things, give them robotic brains, and turn them loose to accomplish thousands of tasks throughout our solar system? What if those gossamer robots carry other robots: crawlers, diggers, crushers, and carriers, and distribute those around the solar system with instructions to support a bold, dynamic civilization spanning across the solar system, with thoughts of going to the stars?

THE PRINCIPLES OF SPACE SAILING

Space sails use solar or other radiation directly as a method of propulsion. They are large, lightweight mirrors which reflect the photons in light or other electromagnetic radiation. Sailing on sunlight or a beam of electromagnetic radiation are forms of photonic propulsion. The advantage with the sails is that a power generator and convertor are not required onboard, thus saving mass and cost. The penalty is the necessity for large sails.

Since photons carry momentum, the reflection of the photons changes their momentum and a resultant force is exerted against the reflecting surface. The resulting force is proportional to the intensity, or power, of the incident radiation. It is inversely proportional to the distance from the Sun in the case of solar radiation. It is also proportional to the square of the cosine of the angle between the sail and the direction of the incident radiation. The action of incident radiation on a perfect sail is shown in Figure 1.1. A perfect sail facing the Sun at a distance of 1 AU will experience a pressure of 9.126 $\mu N/m^2$. In practice, about 90% of that value will be achieved.

The sail loading is an important parameter in determining the performance of a sail vehicle. It is simply the total mass of the vehicle divided by the area of the sail, giving the areal density. It is expressed in grams per square meter (g/m^2) and represented by the Greek letter σ. Sail loading determines the acceleration that the vehicle can experience.

The characteristic acceleration, a_c, is the rate at which the vehicle would

PRESSURE

$$P = 9.126 \cos^2\theta / R^2, \quad \mu N/m^2$$

SAIL LOADING

$$\sigma = \text{Total Mass / Sail Area}, \quad g/m^2$$

CHARACTERISTIC ACCELERATION

$$a_c = 8.12 / \sigma, \quad mm/s^2$$

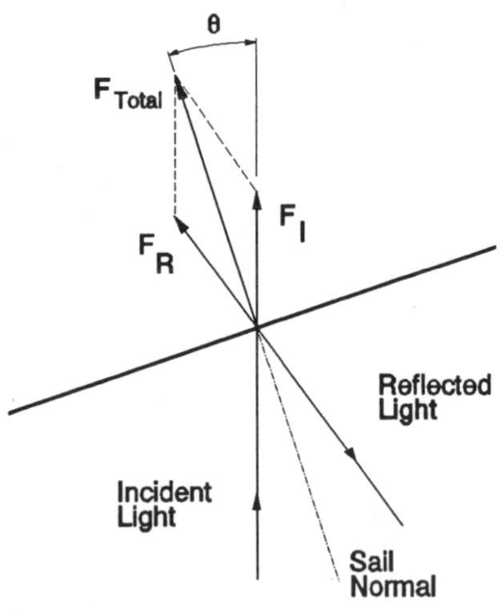

Figure 1.1 The Force Vector of an Ideal Sail.

accelerate at Earth's distance from the Sun if the sail were directly facing the Sun. A related parameter is the lightness number, λ, which is the characteristic acceleration divided by the gravitational acceleration of the Sun at Earth's distance from the Sun. Early sail vehicles are expected to have lightness values substantially less than one, but advanced high-performance sails might eventually reach values greater than one.

A sail does not reflect all incident light. Some of the light is scattered, some is absorbed, and some is transmitted through the reflective surface. These effects reduce the sail performance somewhat below that of a perfect sail. In addition, the absorbed light deposits energy in the sail film. This energy must be radiated back into space if the sail is to be held at a steady temperature. To accomplish this, the back side of the sail is made to work as a radiator to reject the energy being absorbed by the imperfect reflective surface.

In operation, the sail must be oriented to point the force in a useful direction. An attitude control system must exist to execute a steering profile that will cause the sail vehicle to follow a trajectory to its intended destination.

The pressure distribution is not uniform across a sail due to curvature (billow) in the sail. This must be accommodated by the control system. In addition to balancing the ship, the control system must also control the pointing direction of the ship by shifting the center of mass or the center of pressure on the sail.

In addition to photons, space is also filled with streaming particles and electric and magnetic fields. Ships can be made to use the fields and particles for propulsion, but often not with the effectiveness of photonic sailers. An exception is in a region of a relatively intense magnetic field combined with a plasma, such as low Earth orbit, where the field can be used for effective propulsion. However, this propulsion system uses a conducting line rather than a sail and it also requires a power source to obtain a propulsive effect. For these reasons, fields and particles propulsion is not considered to be sailing, at least in this book.

The solar wind is composed of electrons, protons, and heavier charged particles ejected from the Sun. The solar wind particles that impact a sail will exert a small force several orders of magnitude less than the pressure from solar radiation. Solar sails travel on waves of photons, not on the solar wind.

SOLAR SAILING

A sailing ship can maneuver through this or any other stellar system making use of the light coming directly from the star or stars in the system. Our solar system is a simple case since it has only one star. The Alpha Centauri system would be interesting to sail through since two or even three stars can be used by solar sailing ships. Sailing could be of importance to the settlers of other stellar systems.

Solar sails can be used to reach any body in our solar system, illustrated in Figure 1.2. If the body is within the inner solar system, the sailing ship can return to Earth or continue on to other destinations. Solar sailing ships can maneuver away from or toward the Sun, and can also change the orientation of their orbital planes around the Sun.

The ships have the ability to leave or enter orbits around the planets and small bodies of the inner solar system. Within the realm of gravitational influence of a planet or small body, the ships can maneuver to reach or depart from specific orbits where stations might be located. The ships can control their orbits well enough to alter the size, shape, and orientation of the orbits. These are obviously important factors in determining the practicality of using solar sails.

The photons from the Sun stream radially outward, spreading farther apart so that the concentration falls as the square of the distance from the Sun. This causes a sail's acceleration to decrease in the same manner with distance. Obviously then, a solar sailing ship performs better while it is kept close to the Sun.

On any particular voyage, the course is determined by the departure and destination points, the characteristic acceleration, and possibly by the thermal capabilities of the ship and sail. On most missions in the inner solar system, swinging in close to the Sun is too time consuming to be effective. However, this is a very effective way to rapidly gain energy for traveling to the outer solar system.

The sunlight acting on a sail results in a component of force continuously acting in the direction radially outward from the Sun, unless the sail is turned edge-on to the Sun. The sail is turned so as to have a force component perpendicular to the solar radius line which will cause a lateral acceleration. The lateral component of force alters the angular momentum of the orbit, while the total force vector affects the energy associated with the orbit.

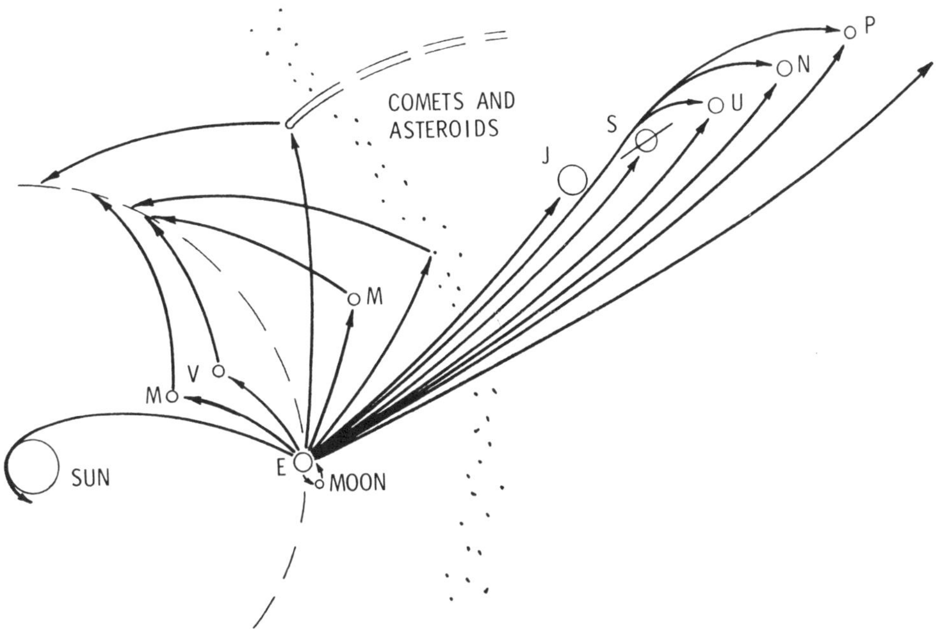

Figure 1.2 The Range of Potential Solar Sailing Applications.

The degree and direction of the turning of the sail controls the subsequent motion of the vehicle. If the sail is turned so that the lateral component of force is against the direction of motion, the vehicle will lose energy and angular momentum, resulting in a spiral in toward the Sun, illustrated in Figure 1.3. If the orientation of the sail results in the lateral acceleration adding to the motion, the vehicle will spiral outward from the Sun.

The sail can be turned so that the lateral acceleration is out of the plane of the vehicle's motion. This will cause the orbital plane to twist around the central body, providing an ability to change the inclination or node of the orbit. Some amount of this type of control is necessary in essentially every interplanetary transfer.

Orbit cranking (twisting) capability improves as the distance from the Sun is decreased. This is because the ship's acceleration increases faster than its

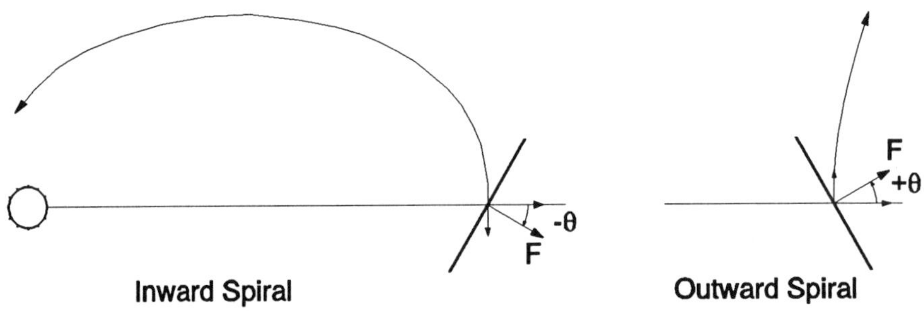

Figure 1.3 Sailing Ships Can Maneuver In Toward the Sun and Outward From the Sun by Setting the Sail Angle.

orbital velocity, so that the sail is better able to turn the velocity vector. Sail missions meant to depart substantially from the ecliptic plane may benefit from going in close to the Sun during part of the trip.

High performance solar sails can open new mission opportunities and reduce the transfer time for other missions. The very high performance solar sails are ones with lightness numbers, λ, of one or greater, thus having characteristic accelerations of at least 5.93 mm/s^2, the solar gravitational acceleration at 1 AU from the Sun. These vehicles can have sail loadings no greater than 1.4 g/m^2.

With a lightness value of one, a solar sail turned squarely toward the Sun would experience an acceleration outward from the Sun equal to the gravitational acceleration toward the Sun. If the sail were kept oriented toward the Sun, the vehicle would move at a constant velocity in a straight line through space, as though the Sun were not present. The vehicle could also fully remove its orbital velocity around the Sun and then hover in a specific position in space.

These high performance ships have very fast transfer times to the planets. They may be able to operate in a practical manner in the more distant regions of the solar system. They can also be used for interstellar voyages in conjunction with laser systems having the capability of generating high intensity beams for pushing the ships.

LIGHT-BEAM SAILING

The spherical spreading of the light from a star limits the usefulness of a solar sailing ship at great distances from the star. However, it is possible to collect light from near the star which would be collimated and directed out to a remote sailing ship by a laser system. A high quality beam would remain nearly constant for a long distance.

The performance of the sailing ship would be enhanced by the use of the directed beam because a greater power density would be reaching the sail. The ship could ride along a stationary beam or the beam could be swept through space, moving along with the ship to get it to its destination.

Sailing with light beams in the outer solar system would require having the beam follow the ship as it in turn follows curving trajectories between the orbits of the outer planets. An interstellar ship would follow a nearly stationary beam that points to where the destination star will be when the ship arrives. Some adjustment in beam pointing would have to be made to allow for the motion of the beam generator around the Sun.

Development of the laser beam system would be a sophisticated, large-scale engineering effort. The return from the effort could be the expansion of our species beyond our home star.

THE INTERPLANETARY CLIPPER

The development of space will be paced in large part by the cost of transportation. Properly implemented reusability is a key factor for achieving low costs. This can be achieved by reusing vehicles that have very low refurbishment needs, and are also inexpensive to operate.

The concept of an interplanetary clipper ship is developed in the following chapters based on the JPL technology and design effort done for a proposed rendezvous with Halley's Comet. An assumption is made that thinner sail films will be available from advances in film processing technology. Another assumption is that less massive structures can be used because of different deployment procedures that can be used in low Earth orbit. A specific type of sailing ship, using a square-rigging concept, is assumed, but this is not a critical assumption since other types can be used.

Sailing ships can move us away from single use, disposable interplanetary spacecraft. They can operate more like marine research vessels outfitted to conduct a wide range of exploratory activities in many different locations.

Solar sail vehicles have a definite potential for providing low cost transportation. The vehicles do not have onboard powerplants for propulsion and they do not consume propellants. There are no massive power generators or engines needing frequent overhaul or replacement, and no massive quantities of propellant needing replenishment after every mission. Solar sailing ships have a potential for operating continuously for two or more decades without substantial refurbishment.

The interplanetary sailing ships can operate with crews or as unmanned cargo carriers. Their operations can support activities ranging from exploration through the commercial needs of established societies across the solar system.

A ship may interface with Earth-based vehicles at an orbital space dock facility. Here outbound and inbound loads exchange vehicles, with the ship taking outbound loads through a spiraling escape from Earth and then entering an interplanetary trajectory.

These sailing ships can carry cargo, robots, and passengers. Yet these are not high-cost, high-technology spaceships. They are simple, reliable ships that can cost less to develop and to operate than the conventional chemical, nuclear, and electric propulsion systems proposed for future space missions.

No other form of space transportation has the potential to operate between Earth and Mars at such low cost. These ships can open up Mars for settlers and commercial activities because of their low costs.

The Mars transportation concepts presently being looked at are the old "use it once and it throw away" concepts, relying on carrying large quantities of propellants. Sailing ships, in contrast, can operate for decades, making numerous voyages in their lifetimes, all without the consumption of propellant.

ECONOMIC COMPARISONS

A simple cost comparison can be made between the use of sailing ships and chemical rockets for delivery of payloads to Mars. Mars mission studies have shown modules of about 20 tons to be a useful size with which to conduct manned exploration of the planet.

The use of chemical rockets requires the development of a new launch vehicle since over 100 tons must be carried into orbit to send 20 tons to Mars. The development cost of the new launcher must be paid, which has been estimated at near $10 billion. The launch rate of this specialized launcher will not be high, resulting in high operating costs per launch and a high amortized development cost. The cost of sending a 20 ton module to Mars using chemical rockets will almost certainly exceed $1 billion including launch cost, amortized launcher development cost, and the cost of the orbital rocket.

When a sailing ship is used, the 20 ton module can be launched by an existing Titan 4 launch vehicle at a cost of about $150 million. The 12 ton, 2 km clipper ship has its unit and launch costs amortized over several voyages to Mars. The development cost is amortized over many ships, each making numerous voyages, so that the accounting charge against each voyage is low. The result is that the cost to send a 20 ton module to Mars by sailing ship will likely be less than $200 million. The time for the module to reach Mars is nearly the same in both cases.

The development, unit, and operating costs of sailing ships are quite low, more in keeping with their mass rather than their size.[1] These costs are low enough for commercial enterprises to cope with, leading to the very real possibility that they will be built and operated by private companies.

REFERENCES AND NOTES

1. Friedman, L.D. (1978) *Solar Sailing Development Program*. Jet Propulsion Laboratory, Report 720-9, Vol. I, Pasadena, California.

2. TO DISTANT SHORES

Solar sailing ships are capable of making a wide range of voyages throughout the solar system. Some are easily conducted as round-trip missions, while others are practical only as one-way trips. In the inner solar system, these ships can conduct round-trips near the Sun, to the inner planets, and to comets and asteroids.

The outer solar system is more difficult to reach. For these missions, the sails can be used to achieve high speeds to minimize the trip time, including the possible use of solar and planetary close approaches to help gain energy. Acquiring the high speeds typically results in gaining escape energy, so the ships will leave the solar system without being able to return. As a ship reaches the destination planet, the payload can separate from the ship to place itself into an orbit around the planet.

Sailing ships can be deliberately sent beyond our planetary system to conduct studies of the extreme outer solar system and interstellar space. They can also be made to travel to nearby stars. Sailing ships driven by powerful lasers are among the most promising candidates for the very difficult journeys to the stars.

MISSION DESIGN

The design of a space mission is based on some objective—a specific payload mass to be delivered to some destination in a particular manner and in a particular time. This involves analysis of both the possible trajectories and the vehicle to reach satisfactory designs of mission and ship. Analysis of the vehicle in the initial stage is a matter of conceptual design and scaling to the appropriate size for the spacecraft. The designs of mission and spacecraft typically undergo many iterations before the final designs are selected.

The development of a mission design begins with a concept of vehicle performance, trajectory data, a payload concept, and some constraints on factors such as mission duration and sail size. This chapter provides a comprehensive summary of the trajectory data which has been generated to date for sail missions, along with some mission concepts based on the trajectory data.

The performance capability of a sail spacecraft depends on the loading of the sail, which is the primary scaling factor in determining the size of the needed sail. The sail loading, σ, is the total mass of the spacecraft (including sail, support structure, and payload) divided by the area of the sail.

Most solar sail missions can be done with a wide range of sail loading values. A heavier payload necessarily means a greater loading on a specific sail and a longer trip time. In those cases where trip time is constrained, then the sail loading also becomes constrained. Analysis of practical trajectories will yield the desired sail loading which becomes a design goal for the sail vehicle. Design and development issues will typically cause some alterations of the desired loading and corresponding trajectory characteristics.

Missions to Mercury, for example, can be done with sail loadings of 50 g/m^2 or greater. A comet rendezvous mission, under a time constraint, may have a loading requirement as low as 6 g/m^2.

If a solar sail were built from currently available materials, the lowest achievable sail loading is expected to be around 5 g/m^2. With some advances in the materials for the sail, loadings of around 3 g/m^2 should be achievable for a sail fabricated on Earth. A sail constructed in space without any need for being folded has a potential for being less than 1 g/m^2, which would result in a very high performance sail vehicle.

TRAJECTORY CHARACTERISTICS

The trajectories available to a sailing ship depend upon the acceleration capability of the ship with its payload. The range of desired departure dates and the characteristics of the departure and arrival orbits are also important.

Optimal Trajectories

The optimization of interplanetary trajectories is usually based on the simple goal of minimizing the transfer time between the departure point and the destination. This means finding the best sail angle for each moment of time to reach the destination in the least time.

The initial conditions at the departure point depend on the speed and direction of motion that the ship can have as it departs from planetary space into interplanetary space, crossing the sphere of influence of the planet. Beyond this zone, the Sun's gravity dominates the motion of the ship. The final conditions at the destination must be compatible with the type of arrival being planned. If the ship is to enter planetary orbit, its speed and direction of motion at arrival must be such that the ship can decelerate into the planetary orbit. The transfer time between planets is sensitive to these initial and final conditions.

When round trip missions are planned, the simple approach of separately minimizing the time for each leg may not give the shortest overall time for the complete trip because of the relative positions of the planets. The optimization of trajectories to multiple destinations can become complex when seeking the shortest total trip time.

Since the ship plus payload mass remains constant along the trajectory, there is no variation in those factors to affect the acceleration. The acceleration along a trajectory varies only with distance from the Sun and the angle of the sail with respect to the Sun. Coefficients representing distance and sail angle effects can be multiplied on the characteristic acceleration to obtain the current acceleration at any point. This makes modeling the dynamic behavior of the ship along a trajectory a rather simple task. (We are not concerned here with modeling the dynamics of rotation—changing the orientation of the sail is treated as a separate problem.)

For operation in planetary orbits, the sail angle must be varied continuously around each orbit. The goal is to maximize the rate of orbit growth or reduction. A simple strategy is to maximize the energy gain in each orbit for escape maneuvers, or to maximize the energy loss in each orbit for capture and descent spirals. The maximization of the energy change per orbit results in achieving close to the shortest spiral times.

In spite of the simple trajectory goals and simple model of the ship, it can be difficult to get optimizing computer programs to converge on the absolute minimum time trajectories due to the peculiarity of the sail force vector in always having a radial component present. There also appears to be a sensitivity to the acceleration value, with characteristic accelerations greater than about 1 mm/s^2 being easier to work with for interplanetary trajectories. Some sophisticated optimization approaches will work with diligence on the part of the analyst, while other sophisticated approaches have not been successful.

It can often be easier to work with less sophisticated approaches toward optimization if there is no demand to find the absolute minimum transfer time. These approaches, which may or may not use the calculus of variations, can provide values which for many applications are reasonably close to the absolute minimum. Details on these trajectory calculation techniques are provided in Appendix C.

Most of the trajectory data in this chapter are based on optimized analyses. Where the data are not fully optimized, they are identified as approximate. In those cases, fully optimized trajectories would be somewhat better than shown.

Orbit Departure and Insertion

Near the end of a spiral escape from a planet, the strategy of maximizing the energy gain per orbit must be modified so that when the trajectory changes from a closed orbit to an open escape trajectory, the outbound speed and direction provide the most benefit for the subsequent interplanetary trip. Numerous variations of spirals and interplanetary transfers must be studied to find the best combination of achievable departure speeds and directions.

A rendezvous of a ship with some target, such as a planet or comet, requires a braking maneuver to remove the velocity which enabled the ship to reach its destination. The approach direction and speed must be such that the deceleration requirements are within the capabilities of the ship for the load it is carrying. Again, numerous variations of interplanetary transfers must be studied to find the shortest transfer time that allows a controlled deceleration into a closed orbit at the destination. Once a closed orbit is attained, the steering profile is selected to achieve the desired size, shape, and orientation of the final orbit. Orbit departure and insertion are discussed in detail in Chapter 6.

Braking by Solar Pressure

It is desirable to use the largest practical force for a braking maneuver, which means having the sail face the Sun during the braking. This is easy at the destination of an inbound trajectory where the ship is naturally moving toward the Sun at the rendezvous. However, the ship is moving away from the Sun to reach other bodies such as Mars from Earth. In these cases, the ship must maneuver to the outside of the body and follow a trajectory that will be approaching the Sun at the time of the rendezvous. This does not make such missions impractical, but it does add some time to the transfer.

Once a rendezvous has been made, the solar force often must be reduced or eliminated for some period of time. To do this, the sail is turned nearly edge-on to the Sun while it is in the part of its orbit where it is moving away from the Sun. This prevents reacquiring escape energy. Returning to the Sun-approaching side of the orbit, the sail is again oriented toward the Sun to continue reducing the orbit size to its desired final value.

An option is to release the payload once orbit is achieved. The ship can then quickly regain energy to escape from orbit and proceed to its next destination. The payload then uses chemical propulsion for further orbit adjustments. This procedure avoids the propellant requirement for planetary capture, which might be the most useful at Mercury.

Aerodynamic Braking

A sailing ship is unlikely to be able to make use of aerodynamic braking. This is because the sail and structure have been designed to take solar force loads only, to obtain the best transfer performance across space. The aerodynamic forces needed for orbit insertion are much larger, and could damage the ship or sail.

Aerodynamic braking may be very desirable for the payload which the ship has been carrying. If the payload can put itself into a planetary orbit by aerodynamic braking or chemical propulsion, the sailing ship can continue on to its next destination without stopping. This increases the utility of the ship since it is not required to spend a long period of time entering and leaving a planetary orbit. An additional savings of time can then be realized since the ship does not have to follow the long trajectory necessary to set it up for entry into a planetary orbit. This allows payload delivery in the least possible time.

Gravity Assist Encounters

A gravity assist, or hyperbolic, encounter is one where the ship will swing around a planet or satellite, using the gravitational pull of the body to alter the direction of motion of the ship. An encounter of this type can add to or reduce the energy of the ship. The encounter can include twisting the plane of motion, which changes the inclination or line of nodes of the orbit.

If the payload has been released at its destination so that it will use aerodynamic braking to enter an orbit, the ship can follow a hyperbolic trajectory around that body to aid it in reaching its next destination.

As an example, a ship can carry a payload from Earth all the way to Jupiter without using any propellant. There the payload is allowed to place itself into orbit while the ship follows a hyperbolic trajectory around Jupiter. Instead of the ship being lost to the depths of space, it will return to Earth for use on other missions.

Payloads meant for more remote planets can be carried by a sailing ship to Jupiter. The payload would be released there on a gravity assist trajectory that swings by Jupiter and takes the payload to its destination. The ship could swing around Jupiter in the opposite direction and return to Earth.

A hyperbolic gravity assist of our Moon can be used to aid departures or arrivals. A ship spiraling out from Earth can direct its trajectory around the Moon on the trailing side, opposite from the Moon's direction of motion. This adds energy to the ship's trajectory which will both reduce the time to escape and give the ship a greater speed relative to Earth when it does break free and enters interplanetary space. An arriving ship would swing around the leading side of the Moon.

The gains from using lunar gravity assists are modest, but may be sufficient to justify the added operating complexity. To use the gravity assists, the departures and arrivals must be timed such that the Moon will be in a satisfactory position, placing a major constraint on the interplanetary trajectory. It is possible that the constraint might cause a time delay greater than the savings resulting from the gravity assist.

SAILING IN EARTH-MOON SPACE

Sailing ships will cross Earth-Moon space on their way to and from interplanetary space. They can also operate between locations within Earth-

Moon space, allowing major cost savings through eliminating chemical propellant consumption. Transportation costs between low Earth orbit and geostationary orbit might be reduced by a factor of 4 to 5 through the use of sailing. The low accelerations inherent in sailing result in transfer times that are long compared to those from using chemical propulsion.

Earth Spiral Trajectories

Sailing ships can follow ascending or descending spirals around Earth, either with the Sun near the orbital plane or almost perpendicular to the plane. Most or all sailing operations in Earth orbit will be near the ecliptic plane since that is the best region for space stations. The perpendicular orbits are necessarily almost polar. It seems unlikely that sailing ships would have any major reasons for operating in near-polar orbits, so this section addresses only low-inclination orbits around Earth.

The times required to escape from various Earth orbits are shown in Figure 2.1 for a range of characteristic accelerations. These figures remain valid for moderately elliptical initial orbits by using the semi-major axis of the orbit as its size. These times are also reasonable approximations of the time for capture and spiral descent to the final orbit.

The data curves are approximations obtained from an analytic approach. These data are thought to be generally valid to within about 20 per cent of achievable values, except for high altitude orbits and high acceleration values which could have larger errors. Exact optimizations require the use of numerically integrated simulations, which to date have not been done in a comprehensive manner.

Departure from a low-orbit space station is represented by the 2000 km curve (the operation is described in Chapter 6). Departure from a geostationary orbit is represented by the 36,000 km curve.

The actual times to escape and the departure speeds are sensitive to the departure direction with respect to the Sun. This is because the component of force radially out from the Sun favors departures in that direction. The departure speed and direction influence the interplanetary trajectory, so optimizing a complete mission trajectory requires studying variations in the escape maneuver. Likewise, the type of encounter with the destination planet must be allowed for, especially if the ship is to enter an orbit there.

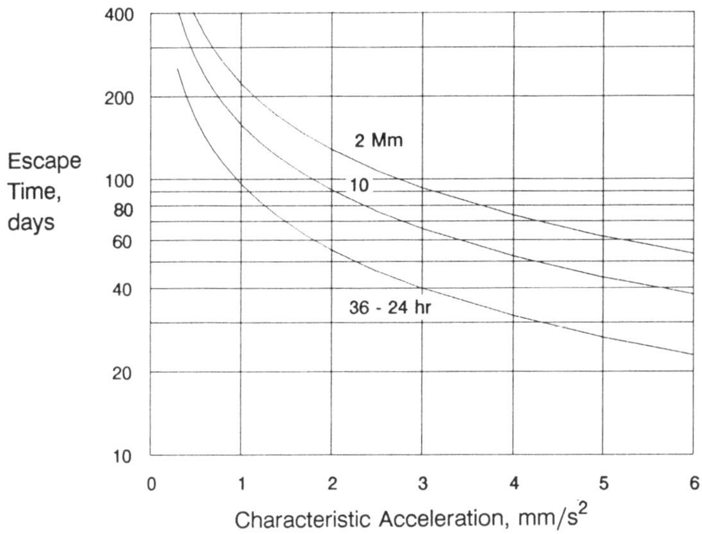

FIGURE 2.1 Earth Escape Times From Various Orbits.

These spiral times can be added to the interplanetary transfer times to obtain an estimate of the total time from Earth orbit to planetary encounter.

In general, the long spiral times characteristic of low-thrust vehicles can make them less attractive than the use of chemical propulsion to move payloads from low Earth orbit into higher orbits. However, the cost of lifting propellant into low orbit is at least $6 million per ton at present. Since sailing ships do not use propellant and they do not have engines which wear out from use, sailing may provide enough of a cost benefit for it to find use someday. Transportation to geostationary orbit is possible, and may be beneficial for commercial use if the transfer times are acceptable.

Lunar Spiral Trajectories

Sail transportation to lunar orbit may be attractive, especially if large quantities of materials are to be carried over a long period of time, as in building up a major lunar base.

Approximate spiral times at the Moon are shown in Figure 2.2. These times can be used as approximations of the time to transfer between the indicated lunar orbits and the L4 and L5 Lagrange points.

Sailing ships can be used as an element in a system for delivering materials mined on the Moon. Packets of material can be launched into Earth-Moon space by electromagnetic accelerators using only electrical energy. Once the packets have slowed from leaving the Moon, sailing ships could capture the packets and deliver them to their destinations.

The ships can also be used to support exploration and mining of material that might be trapped at the Earth-Moon Lagrange points. The reflected light called gegenshein may be an indication that material is held at the Lagrange point beyond the Moon.

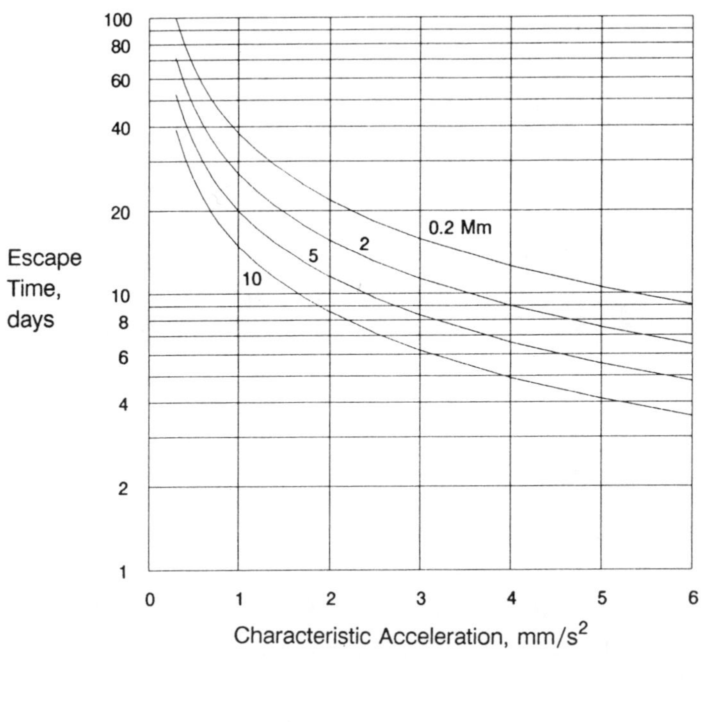

FIGURE 2.2 Lunar Spiral Times.

Levitation

Spacecraft operating in the 24-hour geostationary orbits might use solar sails to provide levitation above the equatorial plane, as illustrated in Figure 2.3.[1] This technique could be used to increase the number of spacecraft operating over a particular Earth longitude. Increasing the number of communication links and observational instruments could provide important commercial benefits.

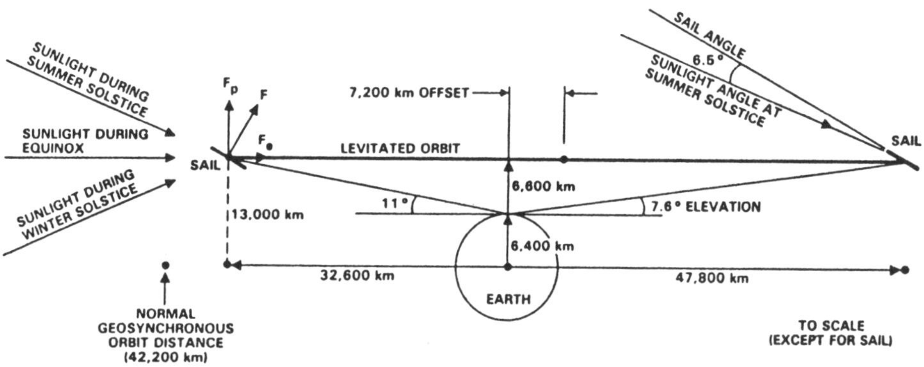

FIGURE 2.3 Levitated Geostationary Orbit. (R.L. Forward/Hughes)

Another mode of levitation uses the sail force to directly counteract gravitational attraction, so that the sail device hovers in space at Earth, another planet, or a small body. This device maintains an equilibrium position at the central body without being in an orbit.

Illumination and Heating

Sails can be used as levitated reflectors in Earth-Moon space which remain over the night side of Earth. The high altitudes at which they would be kept will result in the illumination being spread over large regions of Earth. Reflectors for local illumination can be placed in low Earth orbits.[2]

The local illuminators will not make a significant increase in the total solar energy reaching Earth, but the high altitude reflectors could be made sufficiently large that they would provide substantial heating of Earth. They could be used to modify a future major cooling trend on Earth. This technique can also be used at other planets. Mars in particular might be made more suitable for people if its temperature is increased.

THE INNER PLANETS

When our civilization expands beyond the Earth-Moon region, it will be to Mars as the first logical place. Mercury will also be a potential place, but one that is dependent on effective, low-cost transportation.

Sailing ships will typically leave Earth with low relative velocities, either directly or after a lunar gravity assist to help increase the departure speed. An exception would be for sails that are deployed after the start of the mission, since these spacecraft would be boosted to escape by chemical rockets that could give the spacecraft a substantial departure speed.

A summary of inner planet trajectories is shown in Table 2.1 for ships of two different sail sizes. In the case of the Mars aerobrake arrival, the fastest speed with which the payload would enter the atmosphere is about 12 km/s. The transfer times for the rendezvous missions are somewhat overstated because of conservative assumptions, while the aerobrake transfer is the minimum time.

TABLE 2.1 Inner Planets Missions Summary.

Sail Size m	*Mercury Rendezvous*		*Venus Rendezvous*		*Mars Rendezvous*		*Mars Aerobrake*	
	days	*tons*	*days*	*tons*	*days*	*tons*	*days*	*tons*
800[a]	600	9.4	200	1.4	400	2.4	131	1.9
	900	19	270	4.6	500	5.2	200	5.2
	1200	28			700	9.4	338	10
2000[b]	600	66	200	17	400	23	131	20
	900	124	270	36	500	40	200	40
	1200	184			700	66	338	70

[a]$\sigma = 5.0$ g/m^2 [b]$\sigma = 3.0$ g/m^2 (excluding payloads)

These two ships, carrying 1.9 and 20 ton loads, respectively, could depart a low orbit spaceport by tether release and escape from Earth in about 130 days or go into a lunar orbit in about 150 days.

General transfer times from Earth to the inner planets are shown in Figure 2.4, based on actual orbital elements of the planets.[3] The departure and arrival speeds were assumed to be zero, which gives slightly overstated transfer times.

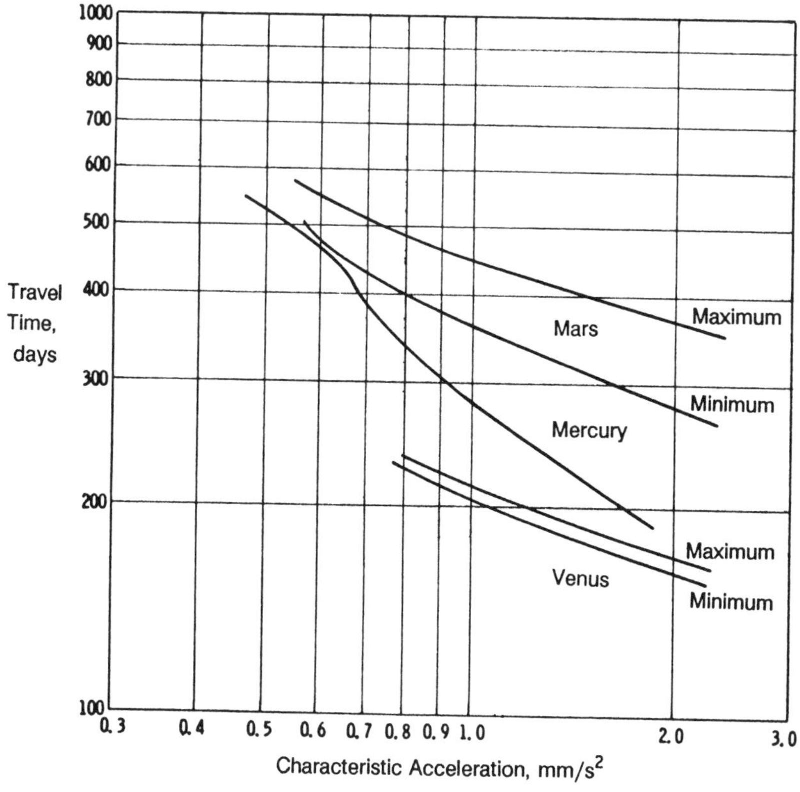

FIGURE 2.4 Typical Travel Times to the Inner Planets.

Mercury

Transportation to and from Mercury has the potential for becoming important to our civilization because Mercury could be the best site for basing the power plants that could drive interstellar ships using certain technologies. Mercury is also the largest source of reasonably accessible iron in the solar system.

Mercury at its closest is only 110 Gm (110 million km) from Earth, a small distance compared to the roughly 10 Tm (10 billion km) from Earth to the edge of our solar system. Yet, starting from Earth, less energy is required to reach the edge of the solar system than to reach an orbit around Mercury. A trip to Mercury is very difficult for ships using propellants; a round trip is out of the question for any such ship in the foreseeable future. In contrast however, a sailing ship can travel to Mercury carrying many tons of payload, and can then rapidly return to Earth after placing its cargo into orbit.

Mercury moves rapidly around the Sun, taking only 88 days to complete an orbit. This provides opportunities every 116 days for inbound and outbound transfers between Earth and Mercury.

In comparing trajectories in Table 2.1, it can be seen that it could be cost effective to carry large loads if time is not critical. A ship could make a 600 day voyage with a particular load, carry three times that load on a 1200 day trip, or five times the load on a 1500 day trip.

The return voyage to Earth is substantially shorter than the voyage to Mercury. With the flexibility in selecting a return opportunity every 116 days, a round trip voyage to Mercury could be done in about two years.

Transfer times to Mercury are shown in Figure 2.5 as a function of characteristic acceleration, based on a departure speed at Earth of 5 km/s. This type of departure requires a chemical propulsion system.

A transfer trajectory corresponding to a characteristic acceleration of 0.85 mm/s^2 is illustrated in Figure 2.6. The departure speed is 780 m/s, and the arrival speed at Mercury is zero. Since the ships can actually remove arrival speeds of 1 km/s or greater, the transfer time of 345 days is somewhat longer than necessary.

Mercury is on average about 0.32 AU from the Sun, which means an increase in the acceleration capability of a ship by a factor of 10 times in the vicinity of the planet. On arrival, a ship can readily sail into orbit around Mercury because of the high solar pressure and the small size of the planet. It can alter its orbit around Mercury including changing the inclination. The

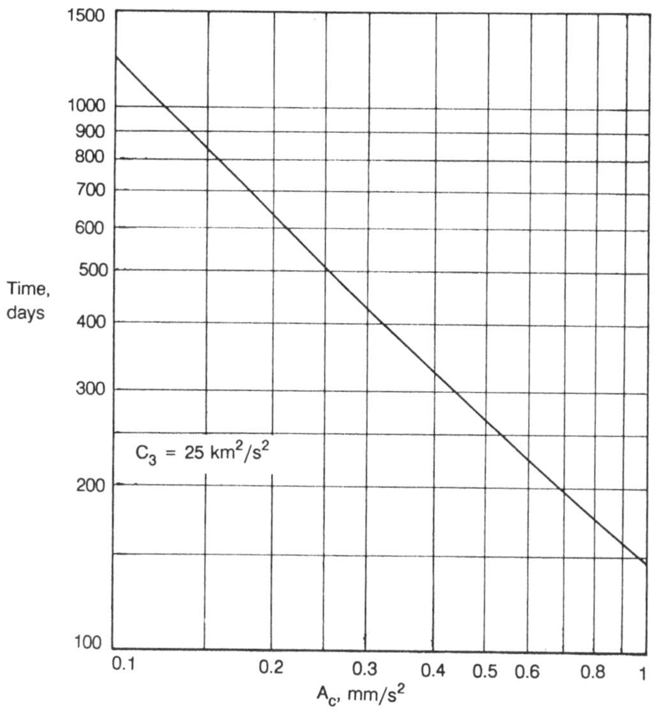

FIGURE 2.5 Transfer Times to Mercury With Fast Departures and Low Accelerations.

ship can also readily break free of Mercury and return to Earth. Escape times from two orbits are shown in Figure 2.7.

Venus

Venus is an object of scientific interest, but it will not likely be important to the development of our civilization for a long period of time. However, sailing ships can easily operate between Earth and Venus, providing scientific returns for relatively low costs. The atmosphere of Venus can be used as an alternative means of braking for the payload, to avoid the time penalty for a ship entering and leaving orbit.

Sailing ships can support observations of Venus, including carrying a permanent observatory into orbit and then maintaining a flow of supplies to

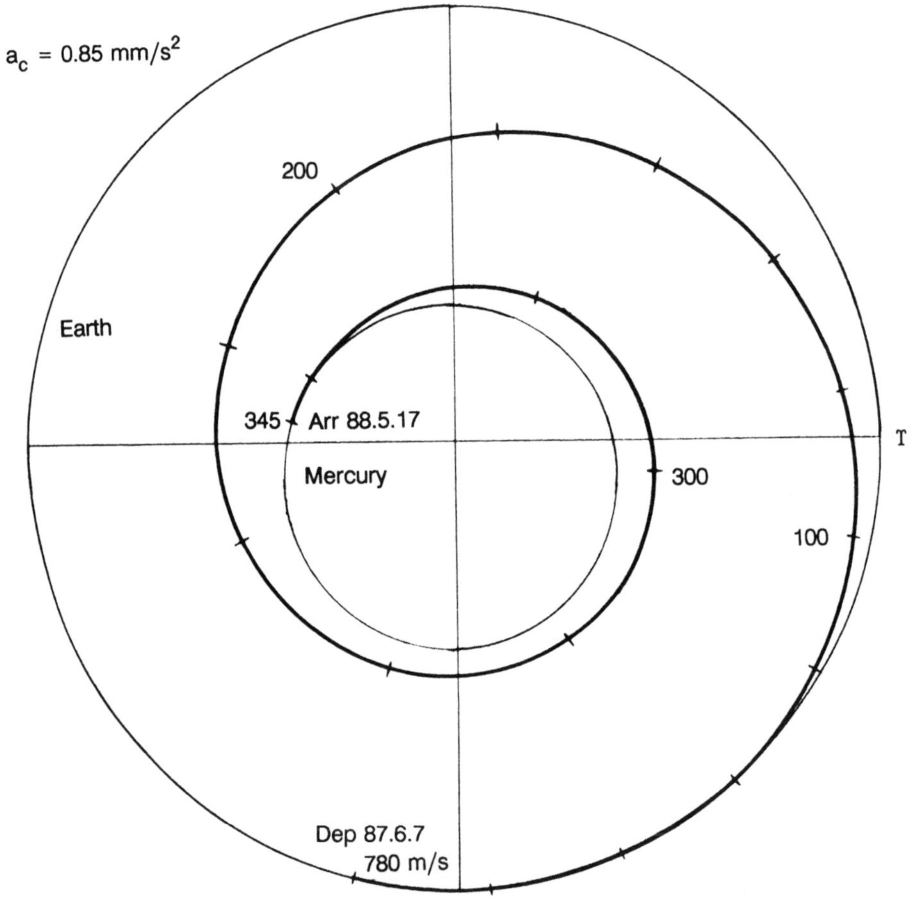

FIGURE 2.6 Trajectory to Mercury from Earth.

it. There may come a time when an effort will be made to transform Venus so that it will support life on its surface. When that is done, sailing ships could support the effort. However, that is very far in the future, and we have no certain knowledge of what transportation technologies will be available then. Sailing ships might be obsolete by that time.

Two transfer trajectories to Venus are illustrated in Figure 2.8. They are based on zero departure and arrival speeds.

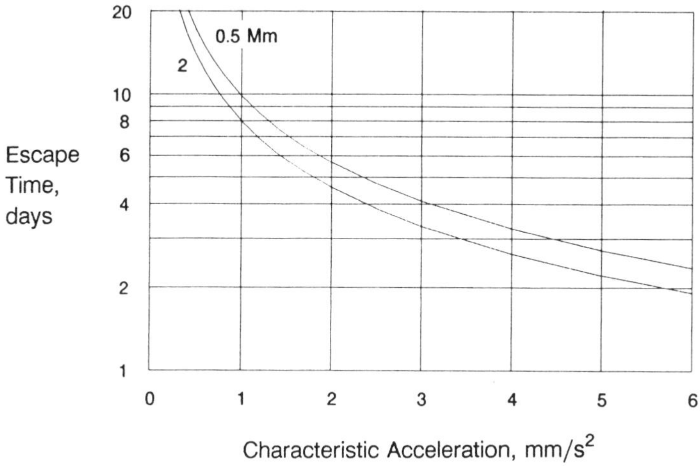

FIGURE 2.7 Escape Times from Mercury.

VOYAGES TO MARS

Transportation to and from Mars will become very important to our civilization. Solar sailing ships could have a major role as part of that transportation system, and this role may be what brings sailing ships into being.

The ships can provide the full range of transportation needs to and from Mars. This includes support of early missions by the delivery of orbital mappers, robotic surface explorers, and sample retrieval spacecraft and landers. At the other extreme of applications, they can transport routine heavy loads and even crew and passengers to support established cities on Mars. The extent of their assignments will depend on costs, risks, and technology developments.

The full reusability of the ships and the lack of propellant consumption provide attractive potential cost advantages. The cost of launching the necessary conventional propellants from Earth are enormous for a manned mission. Use of sailing ships could potentially save more than $10 billion in mission costs.

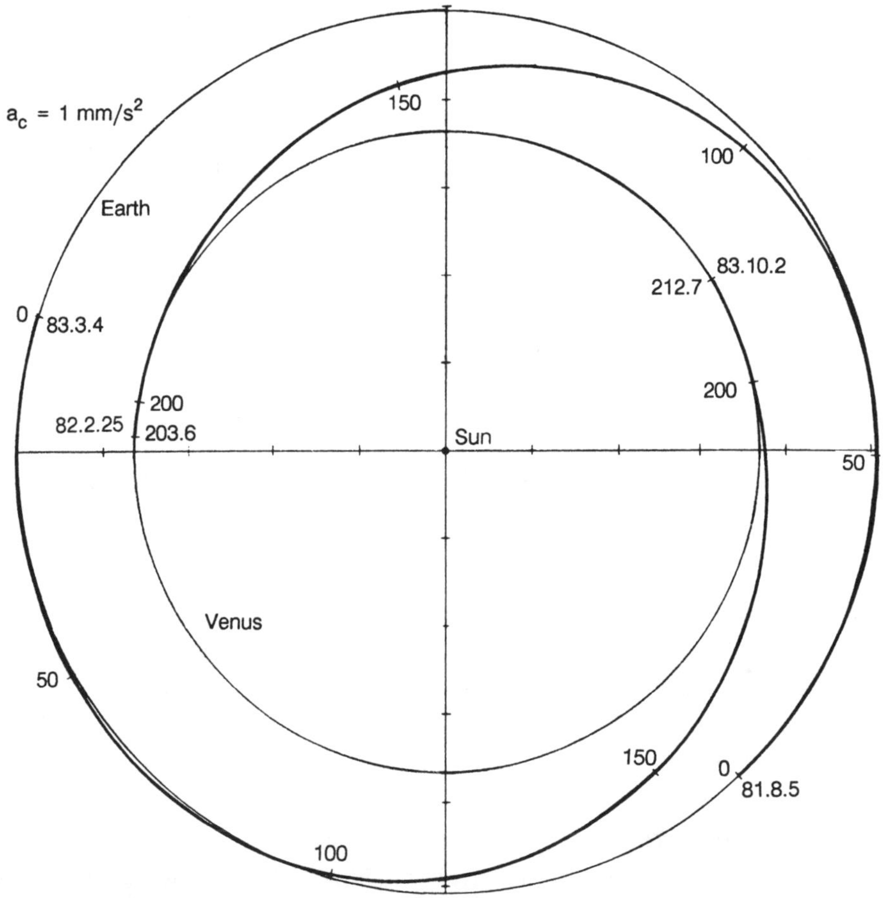

FIGURE 2.8 Trajectories to Venus.

Earth To Mars Transfers

Sailing ships will be leaving Earth orbit at the start of their trajectories to Mars, and they will be entering Earth orbit upon return. At Mars, they will either enter orbit, or they will swing past Mars for return to Earth. These two types of encounters with Mars have a major impact on the time to transfer between the two planets, which will significantly affect the costs of operating the ships and the costs of developing stations and cities on Mars.

The first type, a rendezvous, involves arriving at Mars with a sufficiently low speed that the sailing craft can place itself into orbit. The second type, a swing-by, involves a faster transfer, with the sailing craft releasing the payload as Mars is approached. The payload then directs itself to an aerobraking maneuver that results in the payload entering orbit or descending to the surface. The sailing craft continues past Mars and returns to Earth without having stopped at Mars.

The best departure points from Earth orbit for sailing craft are the L4 and L5 Lagrange points located at lunar distances from Earth. These are the only high Earth orbits with long term orbital stability, where it would be feasible to locate a docking facility. The time required to leave Earth orbit from these locations is small, typically 12 to 13 days. The craft will cross the sphere of influence into interplanetary space with a velocity of 1300 to 1500 m/s at angles of about 50° or greater relative to Earth.

From Earth to Mars Rendezvous

Typical times required to reach Mars for a rendezvous are shown in Figure 2.4 for trajectories based on a zero departure speed at Earth and a zero arrival speed at Mars. In practice, the ships will have significant speeds at both departure and arrival, allowing the transfers to be about two to three weeks shorter than the times shown. These are general trajectories, not tied to specific years.

Two rendezvous trajectories are illustrated in Figure 2.9. They are for launch dates in 1986 and 1992, which are representative of the recurring opportunities. Rendezvous trajectories generally have their best points of departure in the range of 3 to 5 months before opposition.

The transfer times are sensitive to the arrival conditions, with true anomalies between 0° and 180° (periapse to apoapse) being more favorable than between 180° and 360° (apoapse to periapse). This can be seen in the respective trip times of 355.7 and 455.4 days for the two trajectories in the figure.

The Earth escape time of about 12 days must be added to the transfer time to get the total sailing time to Mars orbit. Orbit capture times at Mars will depend on the desired terminal orbit.

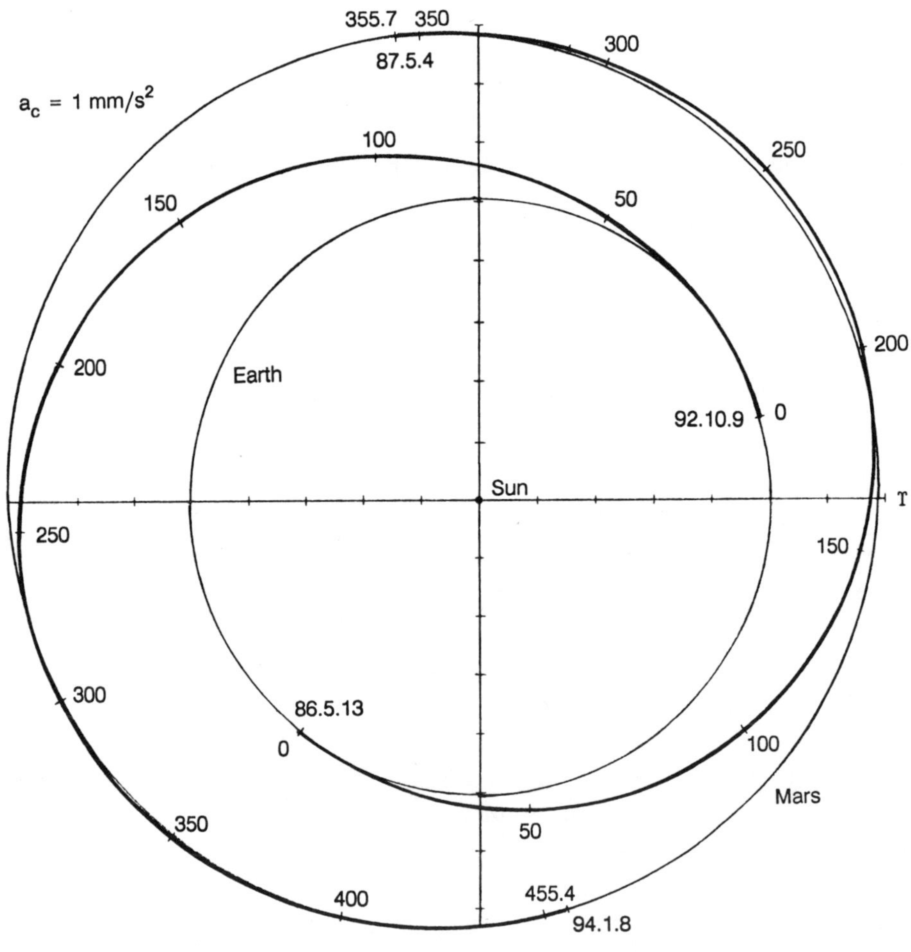

FIGURE 2.9 Rendezvous Trajectories to Mars for 1986 and 1992. A_c = 1 mm/s^2; departure and arrival speeds are zero.

From Earth to Mars With Swing-by

Trip times to Mars are substantially reduced when aerobraking is used to place the payload into Mars orbit, because the ability of the sailing craft to place itself into orbit is no longer a constraint. The sailing craft continues past Mars on a swing-by trajectory.

The envelope of transfer times and entry speeds for these fast trajectories is shown in Figure 2.10. The entry speeds are referenced to an altitude of 120 km at Mars. The transfer times in the figure include 11.6 days spent reaching the sphere of influence from lunar distance from Earth (L4 and L5 points). Departure speed is 1300 m/s, at a departure angle of 60°. The set of launch opportunities between 1998 and 2015 is shown on the closed curve.

It can be seen that, for a_c = 1 mm/s^2, the best transfer times range from about 4 to about 6 months. These times are a function of Mars' position in its orbit when the sailing craft arrives. Entry speeds range from 8 to 12 km/s, also as a function of the encounter location (measured as true anomaly).

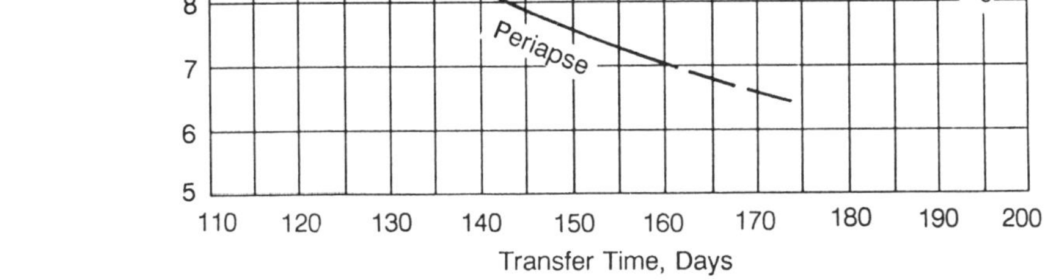

FIGURE 2.10 Transfer Times to Mars With Entry Speeds for Aerobraking Trajectories. A_c = 1 mm/s^2.

The closed curve is the locus of fastest transfers for all possible arrival locations at the orbit of Mars. The lower half represents true anomaly locations between 0° and 180°, while the upper half is for 180° to 360°, moving counter-clockwise. The left extreme corresponds to perihelion arrival, and the right corresponds to aphelion arrival.

The fastest transfer is one with a perihelion arrival, taking 124 days with an entry speed of 10 km/s. The entry speed can be reduced by departing Earth earlier, still encountering Mars at perihelion. This trade is shown by the curve labeled *periapse*. A similar trade can be made for an aphelion arrival, the curve labeled *apoapse*, and all other points on the closed curve. Thus the 2014 opportunity need not have an entry speed of 12 km/s if a longer transfer can be tolerated; launching two weeks earlier would decrease the entry speed to roughly 10 km/s. These trades allow all entry speeds to be held to about those experienced by the Apollo capsules on their returns to Earth.

The best departure dates for these swing-by trajectories are about 60 to 70 days before oppositions.

Mars to Earth Transfers

The best trip time to Earth from near Mars apoapse with a_c = 1 mm/s^2 is 344 days, with a resulting hyperbolic arrival speed of about 1 to 2 km/s. Departure near Mars periapse gives a best transfer time of 400 days, with a similar arrival speed.

These speeds are sufficiently low that the sailing craft can place itself directly into a closed orbit when it reaches Earth. It may be desirable for the crew or payload to separate from the sailing craft and use an aerobraking craft to rapidly achieve a low Earth orbit.

The best return opportunities occur about every 2.1 years. Each of these opportunities will have a transfer time generally between about 344 and 400 days, with an arrival speed that allows the sailing craft to enter a closed orbit.

Round Trips Using Swing-by Profiles

A mission scenario using a swing-by trajectory is shown in Figure 2.11. The departure from Earth is in the year 2003, at a speed of 1300 m/s. The ship arrives at Mars four months later, with the orbiters and landers using aerobraking to enter orbit. The sailing ship continues on, returning to Earth less than two years after its departure. If another sailing ship is used for the return to Earth, the total mission duration is about 29 months.

The ship was assumed to have an outbound a_c of 1 mm/s^2 and a return a_c of 1.7 mm/s^2. The outbound trajectory reaches Mars at a true anomaly of about 90°. The transfer time is 139 days in interplanetary space, which is 151 days including the escape time from L4 or L5. The payload enters the atmosphere at 8.2 km/s for either orbital insertion or direct descent to the surface.

The sailing craft continues past Mars to an aphelion of 2 AU to allow proper phasing with Earth for the return. The arrival conditions for this sailing craft at Earth can be traded between arrival date and approach speed, with a later arrival having a lower approach speed. The return trajectory in this case has an arrival speed of about 1500 m/s, which is an acceptable for reliable capture.

A key factor is that the returning craft enters Earth orbit about 700 days after departing L4 or L5. Since the synodic period is 780 days, about 3 months are available in which to prepare the ship for its next trip to Mars. Thus a sailing craft is able to make a voyage to Mars at every good opportunity, helping to make sailing extremely cost effective.

Meanwhile, the crew or automated equipment that reached Mars after 5 months of sailing is able to operate at Mars for many months. The return time is flexible, but with a departure point that minimizes the sailing time. For this example mission, the stay time at Mars is about 13 months if the return is to be made by a sailing craft that has been based in Mars orbit.

The return trip, also shown in Figure 2.11, takes about 12 months, with the sailing craft able to enter orbit at Earth, with or without its payload. The payload could place itself directly into low Earth orbit using aerobraking.

The ship carrying the crew or equipment from Mars to Earth may arrive near the time when it could sail back to Mars and place itself into orbit in time to carry crew or cargo to Earth at the next good return opportunity.

Round Trips Including Mars Orbit

Some voyages, such as the preceding example, might use the ships in spiral orbits around Mars. These ships must arrive with low speeds to allow orbit capture, and then spiral in and out of orbits at Mars.

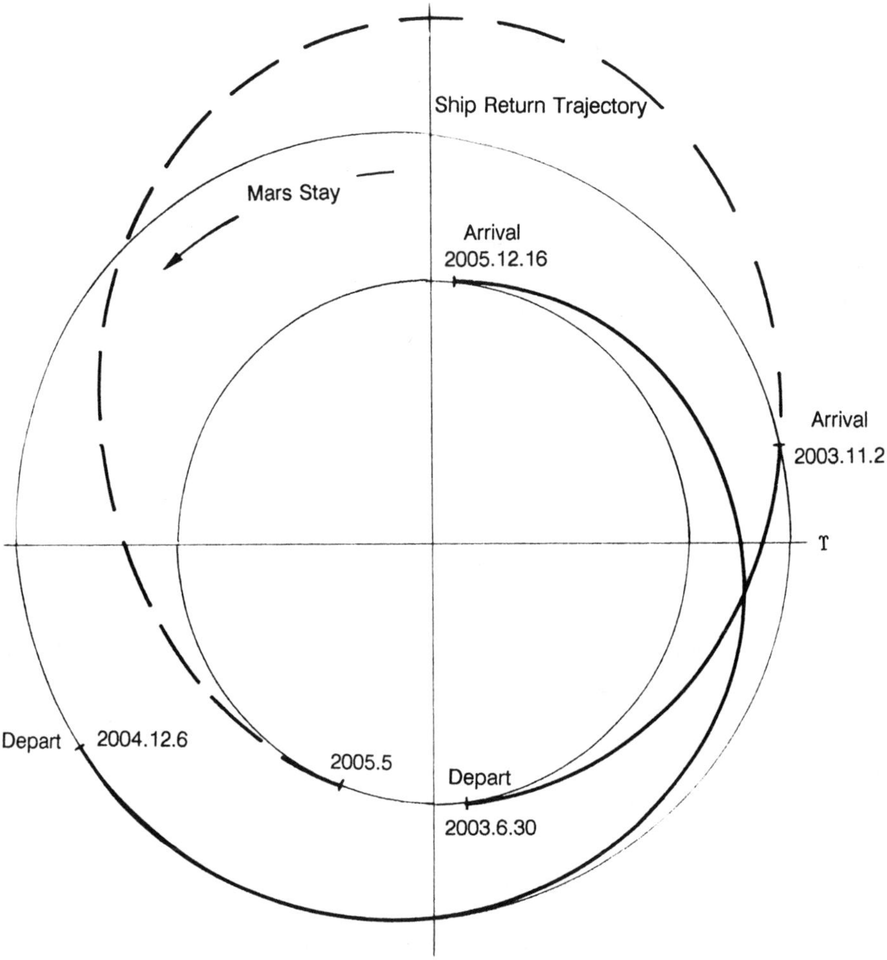

FIGURE 2.11 Round Trip Profile to Mars Using a Fast Transfer in 2003.

Approximate spiral times are shown in Figure 2.12, based on the same analytic approximations used in estimating spiral times in Earth orbit. The departure orbits shown include those of the satellites Phobos and Deimos, plus an intermediate orbit.

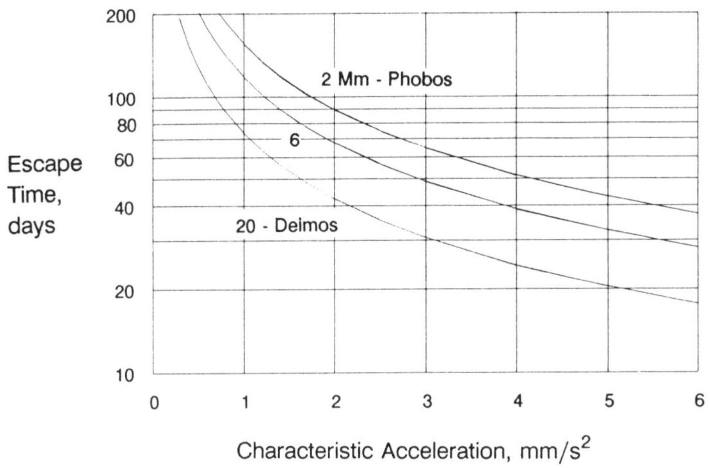

FIGURE 2.12 Spiral Times at Mars.

Special Trajectories

The special trajectories considered here are those associated with non-optimal departure times. These trajectories are of interest for possible early returns from Mars, or special shipments to Mars between good opportunities, as examples.

Non-optimal departures can always be made with sailing ships, but the sailing times can increase substantially. The times can increase to the point of impracticality or loss of benefit.

Early departures, prior to the best time of departure, result in longer sailing times, but typically with reduced arrival speeds in the case of swing-by trajectories. The *periapse* and *apoapse* curves in Figure 2.10 provide quantitative data on these special trajectories using early departures.

Late departures for Earth to Mars trips can create a phase problem because the ship would likely be running substantially ahead of Mars. In this case, the ship will need to go outside of Mars' orbit where its angular rate will be less than the planet's, allowing a good phasing condition to be established.

The ship will then approach Mars in a sunward direction, and can either make an easy insertion into orbit or make a low-speed swing-by.

Late departures from Mars to Earth create more of a problem because Earth will move away from the ship. One approach to the problem is for the ship to cut inside of Earth's orbit, relying on the higher angular rate there to catch up with Earth. The problem with this approach is that the central angle traversed inside of Earth's orbit can easily exceed 360°. This type of trajectory may provide an advantage in arrival date, but it will entail a longer sailing time than simply waiting at Mars for the next good return opportunity.

The other approach to a late Mars departure is more effective for ships with high characteristic accelerations. These ships can sail away from Mars in a retrograde and outward direction, reducing their angular rate below that of Mars, so that Earth will come into position sooner than if the ship had waited at Mars. In essence, this turns a late departure into an early departure in the next cycle. This approach can gain an earlier arrival date, but at the expense of a longer sailing time.

THE OUTER PLANETS

The outer planets can be reached by sailing ships using direct solar energy and using light beams sent from the inner solar system. Even Jupiter, the closest outer planet, is so distant from the Sun that solar sailing is not very effective in its neighborhood. Solar sailing ships have to gain most of their energy in the inner solar system to reach the outer planets. Since the solar energy intensity is so low at the outer planets, solar sails are not adequate for use in orbit insertion at the planets.

To keep trip times from becoming excessive, the ships will have to gain solar system escape energy or close to it. This will prevent the return of ships from the most outer parts of the solar system, but ships going to Jupiter may be able to return by using an energy reducing gravity assist at the planet.

The performance of ships driven by light beams can be drastically better than that of solar sailing ships. With extremely fast transfers, the associated high arrival speeds could preclude entering an orbit at the planet or making a return to Earth, although the payload can be placed into orbit at the planet. Alternatively, the ships could follow somewhat longer trajectories and then go into orbit around the planets using the light beams or return to the inner solar system. For the outermost planets, the return of the sailing ships will probably remain impractical because of the long return time. Even with a light beam available, a sailing ship cannot move toward the Sun faster than the solar gravitational pull allows, unless the direction of the beam is toward the inner solar system, as from a remote reflector.

The method of operation by solar sailing is the same for all of the outer planets as far as the ship's payload is concerned, except possibly when the ship might return to Earth from Jupiter. These payloads are actually independent spacecraft being carried along by the sail. Instead of thinking of the sail as part of a ship that carries various payloads, in these cases the sails might be modules attached to the spacecraft, which are able to execute control over the sail. A complete sailing ship would be used going to Jupiter if it is to return to the inner solar system.

At some point on the trajectory prior to reaching the planet, the spacecraft will jettison the sail module, or the spacecraft will be released by the ship, as the case may be. The spacecraft will trim its trajectory so that it will arrive at the proper point, and then will insert itself into orbit around the planet. The spacecraft will use its own propulsion system for these and subsequent maneuvers.

Arrival speeds are related to the transfer times to the outer planets. Decreasing the transfer time to a planet results in an increase in the arrival speed, and a corresponding increase in the propulsive requirements for the spacecraft placing itself into orbit. For a given characteristic acceleration, there is a minimum transfer time to a planet. The arrival speed for the minimum-time transfer might be too great for a spacecraft design, so there is interest in the variation of transfer time and arrival speed.

Trajectories to the outer planets are indirect except when the spacecraft is started with a chemical rocket or when large characteristic accelerations are used. The indirect trajectories have an initial orbit around the Sun before gaining the energy necessary to reach the destination.

The minimum transfer times to the outer planets, except Pluto, are shown in Table 2.2. The trajectories are based on a characteristic acceleration of 1 mm/s^2, with a zero departure speed from Earth. The table also shows the hyperbolic arrival speed (V_{hp}) at the planet, the number of revolutions made around the Sun, and the size of the initial solar orbit. In these trajectories, the ship makes a moderately close approach to the Sun before going to the planet.[4] An 800 m ship could carry 2 ton loads on these trajectories.

TABLE 2.2 Minimum Transfer Times to the Outer Planets.

	Jupiter	*Saturn*	*Uranus*	*Neptune*
Time, years	2.03	3.29	5.79	8.48
V_{hp}, km/s	16.85	18.50	19.67	19.84
Revolutions	1.32	1.44	1.48	1.52
Apoapse, AU	1.22	1.55	1.93	2.12
Periapse, AU	0.45	0.44	0.43	0.44

$a_c = 1$ mm/s^2, $C_3 = 0$

Jupiter

The minimum transfer time to Jupiter for $a_c = 1$ mm/s^2 is 2 years, as shown in Table 2.2, based on a multi-orbit transfer of about 480°. A direct transfer, requiring less than 360°, is possible, but it requires more time than the indirect trajectory that makes a close approach to the Sun. Slower trajectories can be used that have lower arrival speeds. The transfer opportunities are periodic, repeating every 13 months.

Saturn

Saturn can be reached by direct and indirect trajectories, with the latter being faster. The best departure opportunities repeat at intervals of less than 13 months.

A plot of a Saturn trajectory is shown in Figure 2.13 for a characteristic acceleration of 1 mm/s^2 and a departure speed of zero. The transfer time to Saturn is 4.9 years using the indirect trajectory. This is a slower trajectory than that shown in Table 2.2, with the benefit of a reduced arrival speed.

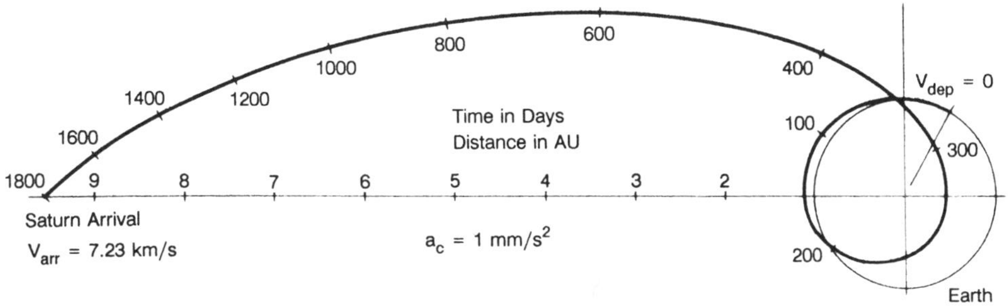

FIGURE 2.13 Multi-Orbit Trajectory to Saturn.

THE SMALL BODIES

Asteroids and comets are called the small bodies of the solar system because they are far smaller than planets or major satellites, but are still significant, observable bodies. They exist in a wide range of types, including stone (silicates), metals, carbon compounds, water, and more volatile gases in their various compositions. They are sources of basic raw materials which could support civilization as it expands into space.

Many of the small bodies are easy to reach by solar sail. Some of these approach or even cross inside of Earth's orbit. Often the near-Earth small bodies have orbits with substantial inclinations from the ecliptic plane, which is Earth's orbital plane around the Sun. However, these inclinations are not a problem for sailing ships because of their lack of propellant consumption. The ships can readily make round trip missions to these asteroids and comets.[5,6]

The small size of these bodies means that their gravitational fields are weak. This allows spacecraft to land on and depart from their surfaces with very little propellant expenditure. The ease of landing means that a spacecraft could leave Earth orbit, travel to one of these small bodies coming near Earth, and land on the surface for less energy than is required to travel to our Moon and land on its surface. The return trip has the same favorable energy requirement.

Inner Solar System Comets and Asteroids

The inner solar system is populated with many asteroids and comets, which tend to be either weak or inactive because of their long exposure to the Sun. Many of these bodies can be reached with modest energy requirements, so that chemical propulsion is a viable option. The bodies that come close to Earth usually have either highly eccentric orbits, often reaching out to the asteroid belt at their most remote point, or they are substantially inclined from the plane of Earth's orbit.

Solar sailing can provide advantages in reaching all of these bodies, but even more so for the highly inclined bodies because of the increased energy requirements for reaching them. An important aspect of using a solar sail is that the ship can return to Earth with samples taken from the bodies. The ship is able to make subsequent voyages to other bodies, enabling a highly cost effective program of investigation to be done with a single spacecraft.

Trajectories to the asteroid Icarus, which makes a close approach to the Sun, are described in Table 2.3. The assumed characteristic acceleration is 1.25 mm/s^2, and the Earth departure speed is zero. The transfer times are shown for several departure years. The sensitivity of transfer time to year of departure is due to the variation in orbit geometry which changes in a cyclical manner, with a new cycle apparently beginning in 1991, where two trajectories are shown. Although the years shown are no longer usable, they provide valid indications of trajectories available in the future.

The comet Encke is nearly extinct, which should allow a sailing ship to safely approach the comet, at least during its quiet times. Landers could probably descend to the surface at nearly any time.

A study was made of the 1982 departure opportunities for Encke. A departure in early April could have had an encounter 594 days later at a distance of 0.75 AU from the Sun, for a characteristic acceleration of 1.03 mm/s^2. This trajectory is illustrated in Figure 2.14.

A trajectory to the comet Giacobini-Zinner is shown in Figure 2.15 for a 1983 departure. The trip time is nearly 7 years for $a_c = 1$ mm/s^2 for this opportunity due to the post-aphelion rendezvous. Rendezvous opportunities should exist near perihelion with significantly shorter trip times in other years.

No systematic study has been made as yet of trajectories to the inner solar system small bodies. This is a task that needs to be done to identify the better opportunities.

TABLE 2.3 Icarus Rendezvous Trajectories.

Launch Date	*Time days*	*Orbits*	R_{min} *AU*	R_{max} *AU*	*Arrival AU*
83-2-28	463	1.36	0.325	1.15	1.90
84-2-24	497	1.37	0.323	1.29	1.85
85-2-15	536	1.39	0.318	1.44	1.80
86-2-07	577	1.41	0.313	1.58	1.74
87-2-01	619	1.42	0.309	1.72	1.70
88-1-27	661	1.43	0.306	1.85	1.65
89-1-21	704	1.44	0.303	1.98	1.62
90-1-17	748	1.45	0.300	2.11	1.58
91-1-13	791	1.46	0.248	2.23	1.55
91-1-12	453	1.50	0.305	1.04	1.91

$a_c = 1.25\ \text{mm/s}^2, \quad C_3 = 0$

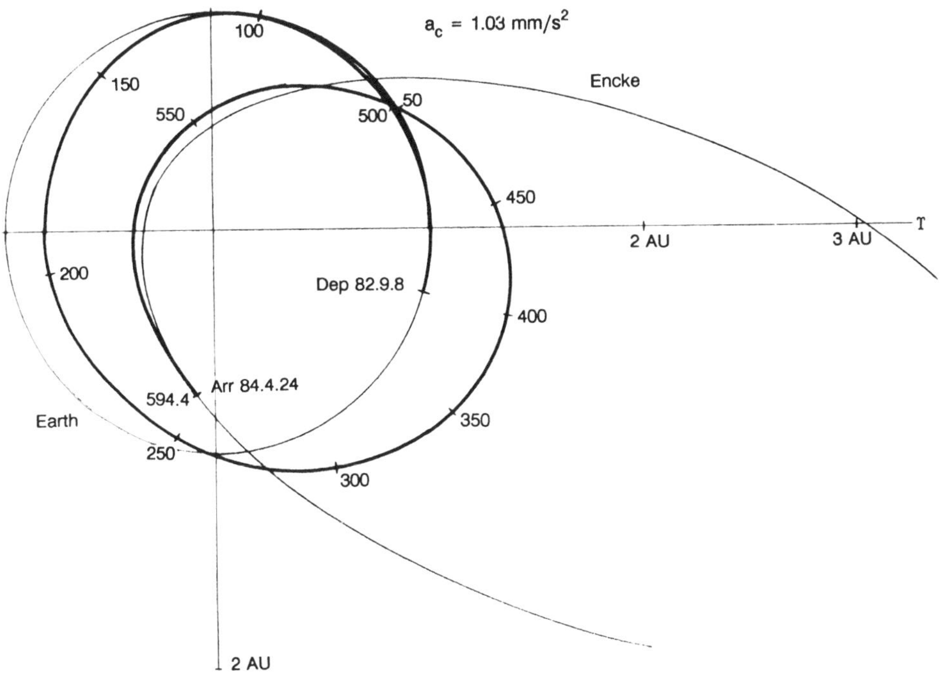

FIGURE 2.14 A Rendezvous Trajectory to Comet Encke.

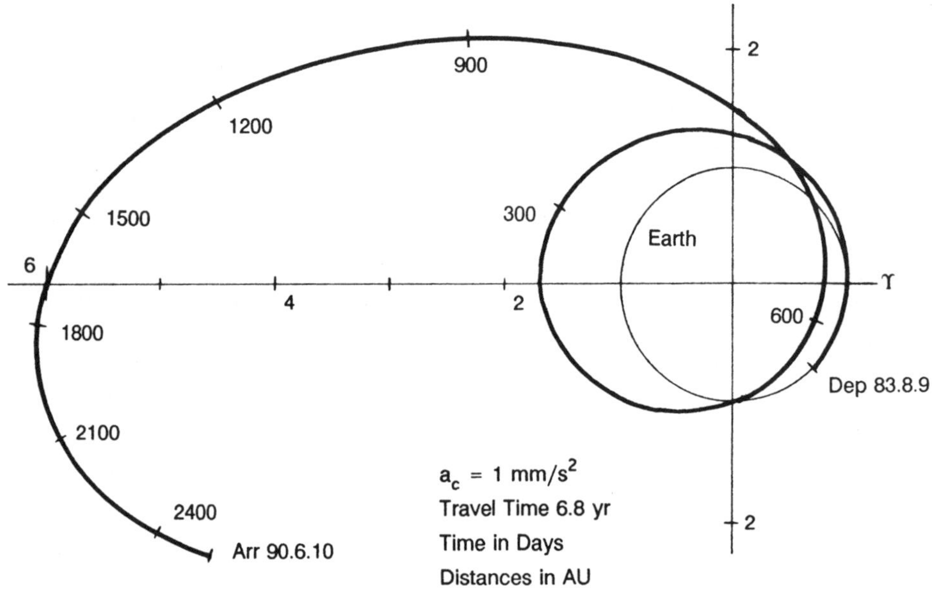

FIGURE 2.15 A Rendezvous Trajectory to Comet Giacobini-Zinner.

Main Belt Asteroids

The main belt of asteroids in the solar system is located near 2 AU from the Sun. At this distance, a solar sail is only one-fourth as effective as it is near Earth. Even so, the sail can move a ship through the belt, exploring many asteroids by entering orbits around them for weeks or months at a time before moving on to the next asteroid.

A multiple-asteroid trajectory was studied for a 1987 departure from Earth. Other multiple-asteroid studies have shown that a good asteroid tour can be started in any year, so this can be considered a representative case. The characteristic acceleration, initially 1.13 mm/s^2, is assumed to increase during the tour as equipment is left at each asteroid visited. The assumed departure speed from Earth is 3 km/s, which is close to what the ship should be able to achieve at departure. A gravity assist at Mars, not included here, may be beneficial for trajectories to the main belt.

The tour, shown in Figure 2.16, begins with a 2.5 year trajectory to the asteroid Vesta. After a 5 month stay, the ship proceeds to Campania on a 3 year trajectory. After a 1 year stay, the ship moves on to Klytaemnestra on a 7 year trajectory. The increasing transfer times are due to the ship moving into the outer regions of the asteroid belt, reaching Klytaemnestra at a distance of 3 AU from the Sun. This study did not include an exhaustive search for rendezvous opportunities after the Vesta and Campania departures.

A tour kept in the inner regions of the belt would have faster transfers between asteroids. As a result, this tour is not necessarily typical. It might be possible to visit asteroids at a frequency more like one per year. Since the ship is not limited by propellant, a survey tour could run for 10 to 20 years.

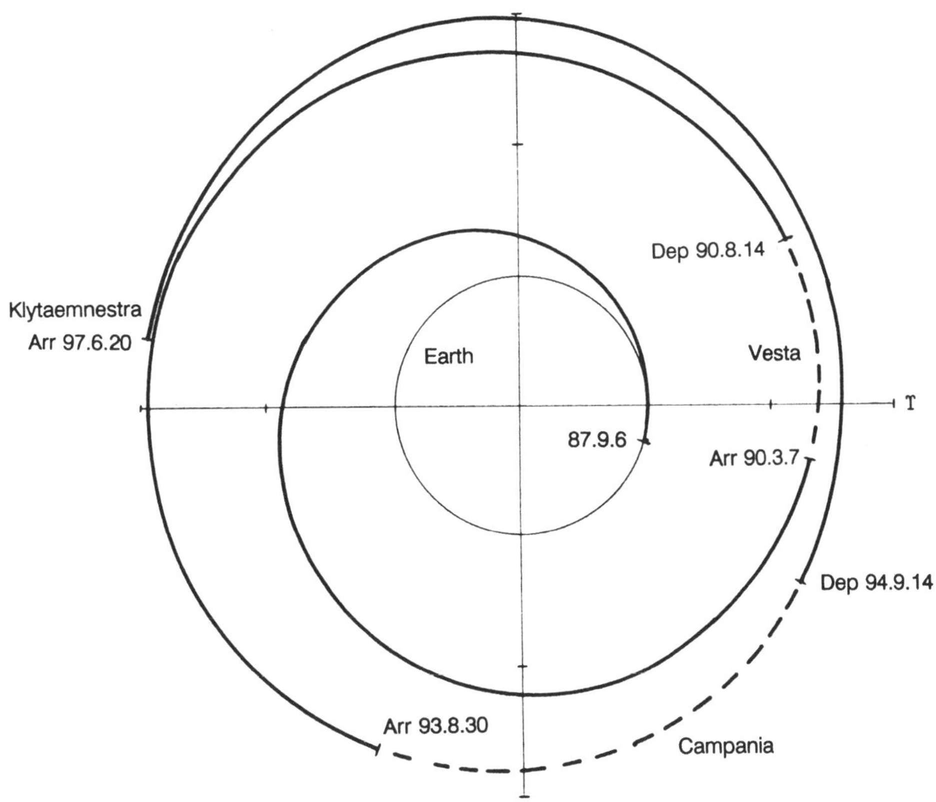

FIGURE 2.16 A Tour of Main Belt Asteroids.

Halley's Comet

A rendezvous with Halley's Comet was looked at extensively for the 1986 passage. The mission analysis was based on this being the first mission for a solar sail, so a conservative assumption on the hardware technology is embedded in the work. The mission data is presented here both for historical reasons and to perhaps serve as a reference point in planning for a rendezvous on the next passage.

The trajectory profile is based on the ship going in close to the Sun to alter the inclination of the ship's orbit. The characteristic acceleration is 1.05 mm/s^2, corresponding to a sail loading of about 7.7 g/m^2. The launch is on a chemical rocket followed by deployment of the sail, giving the ship a speed of 3.5 km/s ($C_3 = 12\ km^2/s^2$) as it breaks free of Earth.

The ship spirals toward the Sun, as shown in Figure 2.17. It takes 234 days to arrive in a circular orbit at 0.25 AU, inclined about 20° from the ecliptic plane.

In this close solar orbit, the ship changes the inclination to about 145°, cranking the orbit across the poles of the Sun in the process, taking about 421 days. This procedure is shown in Figure 2.18.

In the next phase of the mission, taking 943 days, the ship leaves the cranking orbit and sails out beyond 2 AU from the Sun, illustrated in Figure 2.19. As this orbit carries the ship back toward the Sun, the ship is running ahead of the comet as the ship adds energy and does the last inclination change to match the velocity of the comet. The ship is in good viewing position as the comet goes through its brightening process and the active period at closest approach to the Sun (perihelion).

The rendezvous takes place shortly after perihelion, 1599 days (4.38 years) after launch. At this point the sail and supporting structure are jettisoned so that the spacecraft can remain with the comet. The spacecraft approaches closer to the comet as the comet becomes less active. After a thorough orbital survey, the spacecraft lands on the surface for direct observations, ending the dynamic part of the mission.

This trajectory profile is conservative for the next mission opportunity. A moderate performance solar sail ship could follow this type of trajectory, but in much less time and with a rendezvous closer to perihelion.

If a very high performance sail is used for the next opportunity, a much different trajectory might be better. This trajectory would have the ship sail

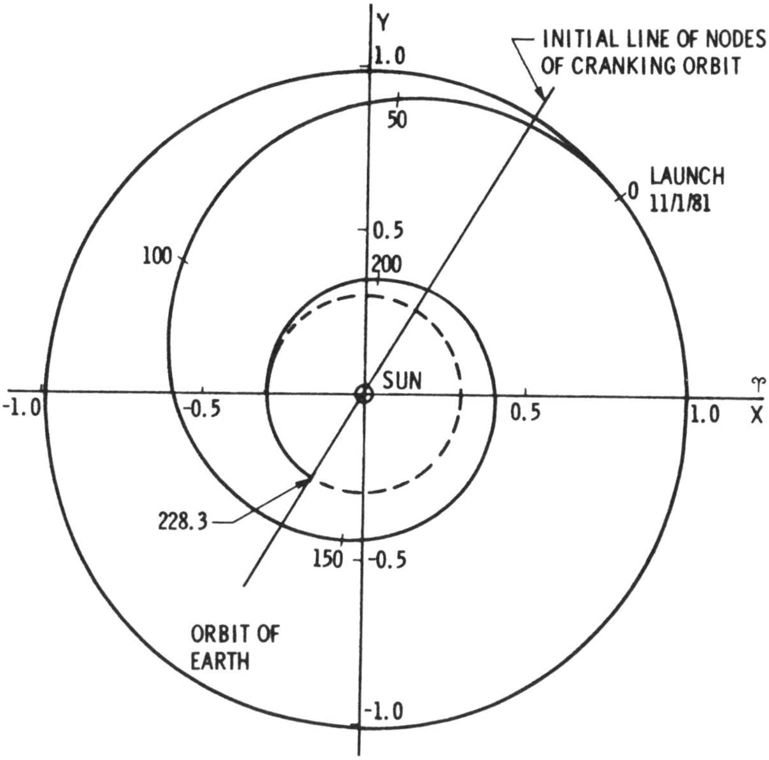

FIGURE 2.17 Halley Rendezvous Trajectory - First Phase. (JPL)

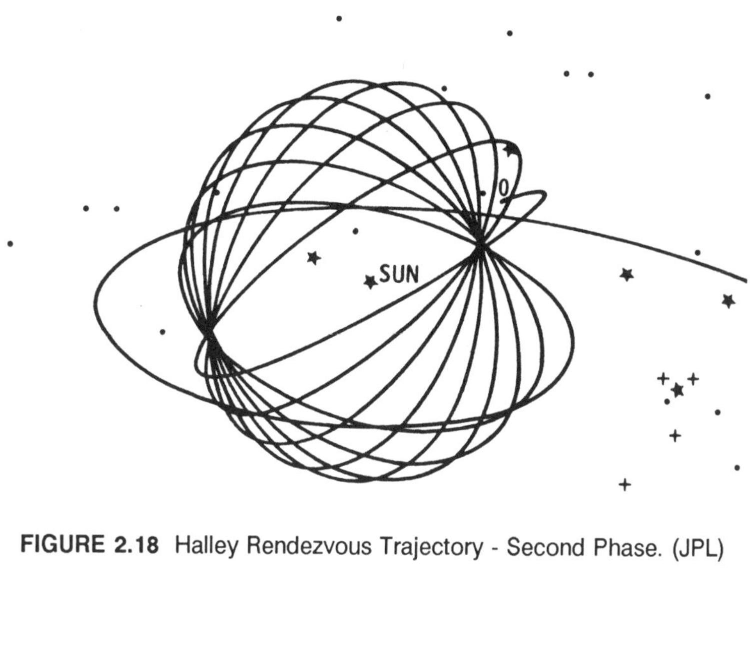

FIGURE 2.18 Halley Rendezvous Trajectory - Second Phase. (JPL)

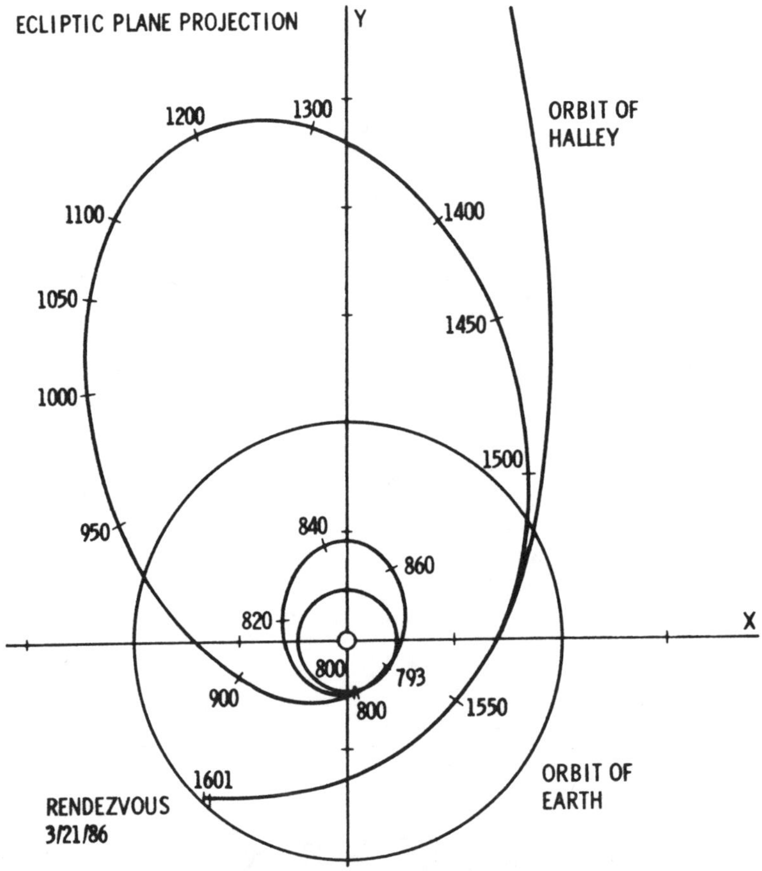

FIGURE 2.19 Halley Rendezvous Trajectory - Third Phase. (JPL)

outward from Earth, reducing and finally reversing its motion through the solar system. The ship gains energy as it falls in toward the Sun, allowing it to rendezvous with the comet near its perihelion. This type of trajectory might be faster than the close solar trajectory, and would have a less severe thermal environment. A full rendezvous with and landing on Halley's Comet is technically and economically possible at the next apparition.

New Comets

New comets are ones approaching the Sun from the outer regions of the solar system, beyond the orbit of Pluto. One of these comets shows up every few years. They are called new comets because they are thought to typically have made only a few passes by the Sun; many are making their first approach to the Sun. These comets are thought to have remained essentially unchanged from the early formation of the solar system. As a result, they are of very high interest for scientific reasons.

When these comets approach the Sun, they have close to the energy needed to escape from the solar system, and they typically have substantial inclinations from the ecliptic plane. This makes the comets difficult to reach in terms of energy requirements. It is also difficult to launch a spacecraft on only a few months notice, but this is essential since these comets are discovered usually a few months prior to their perihelion passage. However, sailing ships with high characteristic accelerations can reach some of these comets on such short notice if the inclination of the comet's orbit is not too high and if the comet approaches from an acceptable direction relative to the position of Earth in its orbit.

If a high performance ship (a_c>5 mm/s^2) were available when a new comet is spotted, an already prepared instrument package and lander could be placed on the ship. Ideally this would be done in a high orbit, such as at L4 or L5, but a tether deployment from a station could be used at the expense of a few weeks spent gaining escape energy.

The ship can reach a perihelion near that of the comet in an estimated 30 to 60 days if Earth is in good position with respect to the comet, otherwise another 30 to 60 days could be needed. The ship can rendezvous with the comet within about 2 weeks following perihelion passage. Necessary changes in the orientation of the orbit are made by the ship as it approaches the Sun and during the perihelion passage. If the geometry between the comet and Earth is very fortuitous, rendezvous can be made shortly before perihelion is reached. In this case, a sample might be returned to Earth.

Detailed observations of the comet's activity can be made during the perihelion passage. As the activity of the comet dies down, the ship can make a close approach for observations, and the lander can be sent to the surface for exploration and analyses of the comet's constituent materials.

SAILING IN SOLAR SPACE

There are many possible destinations in solar space other than planets and small bodies. These include the Lagrange points of the planets and various orbits and points around the Sun for studying and monitoring its activities.

Sun-Earth Lagrange Points

The Sun-Earth Lagrange points may hold small bodies that are of interest for study and mining. Sailing ships can reach all of these points, and could be the most cost effective means of doing so.

The Lagrange point located between the Sun and Earth (L1) is easily reached by sailing ships. The energy requirement for transfers are low, but the transfer times could be several weeks to a few months.

There is a potential important application of sailing technology at locations slightly closer to the Sun than the L1 point, but which remain synchronized with Earth as it orbits the Sun. Sails, acting as shields or concentrators, can be placed facing the Sun in equilibrium locations, or in halo orbits around the locations, and held there indefinitely if periodic maneuvering corrections are made. The equilibrium position is inside of L1 due to the solar pressure on the reflector.

The sails will reflect away some of the sunlight which would otherwise reach Earth, or can increase the incident sunlight.[7] The reduction in incident solar energy could potentially offset expected temperature increases resulting from the buildup of the greenhouse effect. Several decades would likely be required to construct a large shield, but the project might be done before extensive damage results from the greenhouse effect. Even if the greenhouse problem is solved on Earth, there is a longer range problem resulting from slowly increasing solar output. The same solution could be used for this problem. This technique could also be used to reduce the temperature of Venus.

Halo orbiters can also be used to increase the temperature of a planet such as Mars by reflecting additional energy onto the planet. Suitable locations are L1, L2, and the region around the planet.

Close Solar Approaches

The energy requirements for trajectories passing near the Sun are even greater than for Mercury, which was previously stated to be more difficult than leaving the solar system. As with Mercury, sailing ships can meet the energy requirements in relatively short periods of time, without consuming propellant, and preserving the vehicle for subsequent uses.

We have an obvious critical dependence on the Sun, but none of the spacecraft used to observe it have approached closely, nor have they observed the polar regions. A spacecraft will make a single observation pass over one of the poles in the 1990's, but it will be roughly 1 AU from the Sun during the passage.

The design effort for the Halley Rendezvous mission provided a sail design which can operate at 0.25 AU or closer from the Sun for at least 1 year without significant damage to the sail or its supporting structure. The technology effort on materials indicated that sails would probably be capable of sustained operation inside of 0.2 AU.

The trajectories in to circular orbits close to the Sun are the same in geometric form as that for the Halley Rendezvous mission. The times to reach various circular orbits are shown in Figure 2.20 for characteristic accelerations of 0.5 and 1 mm/s^2.

Sailing ships with only moderate characteristic accelerations can be used to drop probes directly into the Sun. This can be done over any point on the surface, including dropping probes straight in along the polar axis. A ship can release a probe and then alter its own trajectory to avoid incineration. A single ship can drop several probes during a mission.

The ships obviously then have the capability of also placing probes into orbits where they will pass as close to the surface as desired, limited by the thermal capacity of the probe or sacrificing the probe in a penetration of the solar atmosphere. The probes can go well inside the distance the sail can tolerate. This can be done without losing the ship since the ships can quickly gain energy after releasing the probes.

The distance to which a sail can operate is dependent on the thermal and structural load capacity of the design. An ordinary sail with a high emissivity can go inside of 0.2 AU, while a specialized sail might withstand reaching 0.06 AU (about 8 million km, or 15 solar radii from the surface).

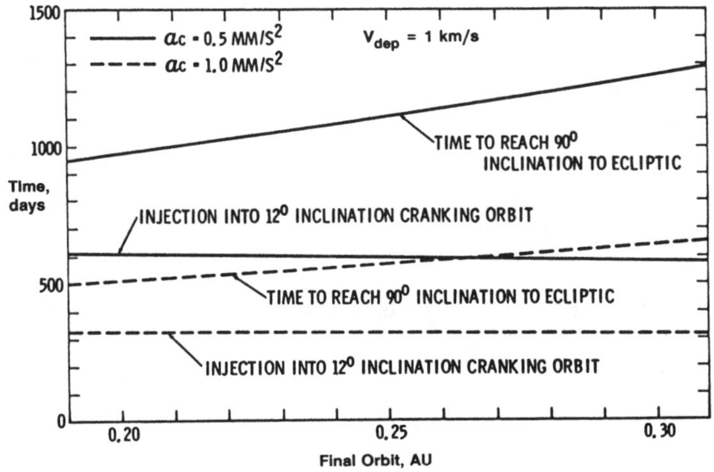

FIGURE 2.20 Transfer Times to Close Solar Orbits. (JPL)

High Latitudes

If the desired solar orbit is inclined to the ecliptic plane, some of the inclination change is achieved during the inbound trajectory from Earth. Any remaining plane change is done at the desired final distance from the Sun, in the same manner as planned for the Halley Rendezvous mission.

The times required to reach an inclination of 90° from the ecliptic plane are shown in Figure 2.20 for moderate characteristic accelerations.

In spite of the extreme energy requirements, sailing ships can reach close polar orbits around the Sun, release an instrument package, and then go on to other tasks or return to Earth. The released package can be left in a polar orbit, or, with its orbital motion removed, allowed to drop as a probe directly into the pole of the Sun. Alternatively, the sail can be kept with the instrument package to continuously alter the orbit.

Longitude Synchronization

An instrument package can be placed in a 25 day orbit to remain synchronized with the rotation of the Sun. If it is placed in the Sun's equatorial plane, the package remains stationary over the same point on the Sun, as shown in Figure 2.21. The energy requirements are so great for this mission that sailing ships are probably the only practical and cost effective means of transportation.

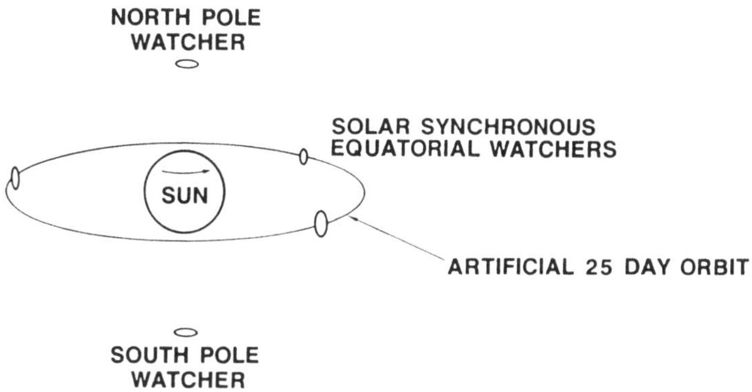

FIGURE 2.21 Synchronous Solar Orbits. (R.L. Forward/Hughes)

If the sail is kept with the package, the observation point can be shifted. The instruments can be stationed over points of interest or moved away from eruptions that threaten the spacecraft. Alternatively, the ship could send probes into eruption zones.

The radial component of force on the sail effectively reduces the solar gravity influence, allowing use of orbits farther removed from the surface if that is preferred. The sail can also levitate the orbit out of the equatorial plane so the spacecraft remains at a fixed latitude. Under this condition, the sail can still be used to alter the longitudinal position while maintaining the desired latitude. The angle of levitation is a function of the characteristic acceleration.

Hovering in Space

A ship with a characteristic acceleration of 5.93 mm/s^2 does not need any circular motion around the Sun to prevent solar gravity from pulling it in. The ship can hover at any point in interplanetary space, balancing the pull of gravity with the pressure of sunlight.

The poles of the Sun are points of special interest for hovering. Ships can hover above the poles under essentially permanent conditions, as close in to the Sun as their thermal capabilities will allow, Figure 2.21. This could probably be inside of 0.2 AU, less than 50 radii above the surface of the Sun.

DEEP SPACE

Deep space as used here refers to the volume of space beyond the orbits of the outer planets. This is a vast region empty of matter except for the rarified interstellar gases, dust grains, cosmic ray particles, and an occasional small body of frozen volatile substances left over from the formation of the solar system. These small bodies are sometimes sufficiently perturbed in their motion that they approach the Sun to become comets, developing the cometary halos and sometimes tails. It is thought that a large number of these cometary nuclei exist, forming what is called the Oort cloud, far beyond the outer planets.

There is some scientific interest in exploring the regions beyond the outer planets. Directly measuring the types and numerical densities of the small bodies, dust grains, and gases can help in understanding the formative processes of solar systems. Exploring spacecraft might even find planets very remote from the Sun.

Solar sailing ships can launch experiment platforms and multiple-spacecraft buses toward deep space. The spacecraft could use chemical or other propulsion to alter their trajectories in deep space, so that the spacecraft become new members of the Oort cloud.

Neptune can be used to aid circularization or change inclination if the spacecraft are not going out much farther than Neptune's orbit. A Neptune gravity assist is usable for more remote regions, but may be prohibitively time consuming.

INTERSTELLAR VOYAGES

Interstellar travel is one of the imperatives for our species. It is an emotional and intellectual imperative to explore, to develop what we find, and to expand our cultural horizons. It is also a physical imperative, because our Sun is destined to perish. Humans will have the choice then of perishing, traveling to other stars, or accomplishing some massive engineering feats to continue living in this solar system. This physical imperative is almost certainly a moot case, since it appears extremely unlikely that humans will not take up interstellar travel sometime within the next few billion years. Interstellar travel might begin before the end of the Twenty-First Century.

Solar sailing can be used for interstellar travel if extremely close approaches can be made to stars. At present, there are no sail concepts that are known to meet the intense thermal environment and have sufficiently low mass.

With an a_c of 50 mm/s^2 (σ = 0.16 g/m^2), a ship passing only 4.4 solar radii above the Sun's surface can achieve a speed of about 810 km/s leaving the solar system. This results in a trip time of 1600 years to reach Alpha Centauri. In this profile, the peak acceleration is 41 m/s^2, slightly greater than 4 g's. With an a_c of 100 mm/s^2 (σ = 0.08 g/m^2), a ship passing 2.6 solar radii above the surface can reach Alpha Centauri in 780 years. These performance levels are overstated because the simulations did not allow for the effects of the solar disk.

Laser sailing ships are among the most promising of the numerous interstellar concepts that have been studied. These ships can attain the speeds necessary to achieve reasonable trip times. They also may be the most cost effective since the driving power system remains in our solar system, and can be used for more than one ship.

The performance of a laser sailing ship is a function of its acceleration capability, which depends on the sail loading and the intensity of the laser beam driving the ship. The time to reach the intended solar system is also a function of its distance from Earth, and whether or not the ship will stop when it arrives. A discussion of the operations is contained in Chapter 7.

A time of 100 years is often considered to be an upper limit on trip times to other stellar systems without going to the extreme of using slow-traveling planetoids instead of ships.

The nearest stellar system, Alpha Centauri, is 4.3 lightyears (ly) distant, so that an average speed of at least 0.043 c (4.3% of the speed of light) is required. The Sun-like stars Epsilon Eridani and Tau Ceti are 10.8 and 11.8 ly distant, respectively. Voyages of 100 years duration to them require average speeds slightly greater than 0.1 c.

These requirements could be met by laser-driven sailing ships if the necessary power levels and areal densities can be achieved.

REFERENCES AND NOTES

1. Forward, R.L. (1981) Light-levitated Geostationary Cylindrical Orbits. *Journal of the Astronautical Sciences*, Vol. 29, No.1, 73-80.
2. The concept of using orbital reflectors for illumination was invented by Hermann Oberth.
3. Sauer, C.G. (1976) *Optimum Solar Sail Interplanetary Trajectories*. American Institute of Aeronautics and Astronautics, 76-792, New York.
4. Sauer, C.G. (1976).
5. Sauer, C.G. (1976).
6. Wright, J.L. and Warmke, J. (1976) *Solar Sailing Mission Applications*. American Institute of Aeronautics and Astronautics, 76-808, New York.
7. The concept of using sails at L2 for temperature control of Earth was invented by the author in 1976. The concept has subsequently been published by several other authors.

3. SAILING SHIP DESIGN

Sailing ships can be made in a wide variety of configurations. The basic requirement is simple: spread a large reflective sheet in a manner such that it provides a controllable force from radiation pressure. The sail can be held in position by stiff structural members or by the centrifugal effects from spinning. With both of these modes, a wide range of potential designs have been identified which appear workable. The actual choices for implementation will depend upon practicality, cost, and performance. The more promising design concepts identified to date are presented in this chapter.

The overall vehicle is referred to as a ship, as it is in maritime usage. What might be thought of as the hull is a minuscule structure compared to the sail. This is the operations module, which houses the operating systems and carries the payload. The sail itself is integral to the ship design in the overall geometry.

DESIGN CONSIDERATIONS

The various configurations of sailing ships can generally by classified among three basic types. These types are the spar-supported sail, the spinning disk sail, and the spinning blade sail, shown in Figure 3.1.

A specific application (mission) might require evaluating all three types as candidates. Some situations, which might require stationkeeping or rapid attitude change capabilities, could immediately eliminate some configurations from further consideration. If orbital facilities are established for the

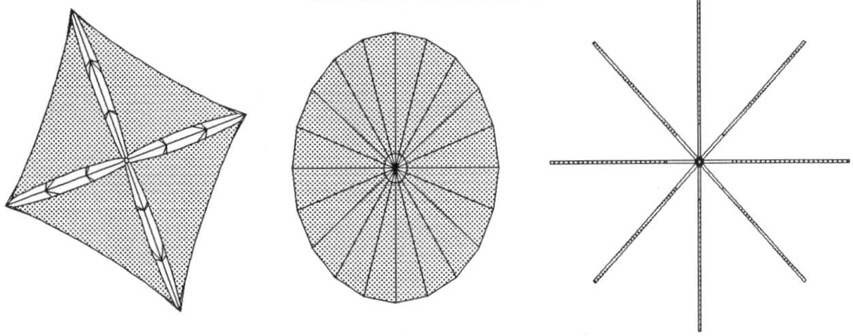

FIGURE 3.1 Basic Types of Solar Sailing Craft.

construction or erection of a type of sail, that type may become the one of choice for subsequent applications for economic reasons.

All of the general system engineering factors inherent in spacecraft design must be considered in designing a sailing ship. The operation of sailing ships is much like that of other spacecraft in terms of operations of the onboard systems. These systems must provide supporting structures, operating mechanisms, electric power generation and distribution, guidance and navigation, attitude control, command and data handling, communications, thermal control, and payload support.

An autonomous ship will include these functions in its onboard systems, making it independent of the load it carries. These systems might be contained in an operations module that remains with the sail until removed for overhaul or upgrading.

The sail and its structures and mechanisms can also be treated as a module attached to a spacecraft. In this approach, the sail module is dependent on the spacecraft for the necessary operating functions. The sail module might be jettisoned by the spacecraft at some location in space when the sail is no longer needed. This approach can be taken when there is no interest in reusing the sail.

The design and operations of the onboard systems are not described in this book where they are conventional in their requirements and design approach. Where peculiarities do exist, they are treated in the discussions of the specific configurations. Even where the systems are basically conventional however, there can be numerous challenges to the designers, especially in placing appendages such as solar panels and antennas.

In addition to these issues, the sail system presents its own special factors, many of which impose requirements on the other systems of the ship. The design procedure must include a determination of the sail size, the sail shape and configuration, and the areal mass loading on the sail.

The pressure distribution must be evaluated in all potential attitudes to determine the control requirements. Control requirements are also determined from maneuvering requirements that come from an analysis of the mission operations. The sail and structural deflections from solar pressure, maneuvering, and gravity gradients must be determined. The thermal response of the sail to solar distance, sail orientation, and planetary reflections must be determined. The pathways for reflected light must be evaluated to avoid possible thermal damage. The deployment scheme must be defined in detail, evaluated, and tested. Many special design factors peculiar to each sail type must also be evaluated, including, for example, the dynamic behavior of spinning sails.

The design effort at JPL in the 1970's included these factors and others, especially the short development time and the need for automated deployment in the Halley Rendezvous mission. This effort showed that solutions or alternatives can be found for design problems, and that viable concepts for sailing ships can be created.

Operating life, maintenance, and repair are important issues for ships that are intended for numerous voyages. The Halley Rendezvous design appears to be capable of operating for at least a decade without maintenance, and longer if it is maintained.

SIZING THE SAIL

The size of the sail needed for a ship is determined by two principal factors: the mass of the payload to be carried, and the nature of the trajectory for the mission. An analysis of the trajectory from the point of departure to the destination for the desired trip time yields the characteristic acceleration, a_c, that is necessary for the mission. This is related to the resultant sail loading requirement by the expression

$$\sigma = \frac{\eta \cdot 9.126}{a_c} \approx \frac{8.12}{a_c} \tag{3.1}$$

where η is the overall sail efficiency, a_c is in mm/s^2, and σ is in g/m^2. The efficiency factor for the Halley Rendezvous design was estimated at 0.88 to 0.90, with 0.89 taken as a representative value.

The required sail area is determined from the total cruise mass of ship and payload divided by the desired sail loading. This may conveniently be expressed as

$$\text{sail area} = \frac{m_{\text{total}}}{\sigma} = \frac{m_{\text{load}}}{\sigma - \sigma_{\text{ship}}} \tag{3.2}$$

when the sail loading of the ship without payload is already known.

The payload is shown as a function of the characteristic acceleration for several values of sail loading of the ship without payload in Figure 3.2. In this figure, the sail-related structure is assumed to be 20% of the sail mass.

CONFIGURATION SELECTION

The factors discussed in this section are those affecting the design of a solar sailing ship meant to operate in the region between Mercury and the asteroid belt. Some of the specifics of laser-driven ships are also discussed. The factors discussed are only the most prominent ones, excluding costs, which can be usefully estimated when specific applications are being investigated.

Sail loading is commonly the predominant selection factor. The different solar sail concepts under consideration are expected to have somewhat differing values of sail loading. However, a surprise turned up in the design effort at JPL in 1977. A spar-supported square sail and a spinning sail called the heliogyro were both subjected to intensive design analyses. These two drastically different designs were found to have sail loading values which differed by only a few per cent, within the mass estimate uncertainties of the two designs. The expectation had been that the heliogyro would have the better performance, but when they are compared on the basis of equal sail

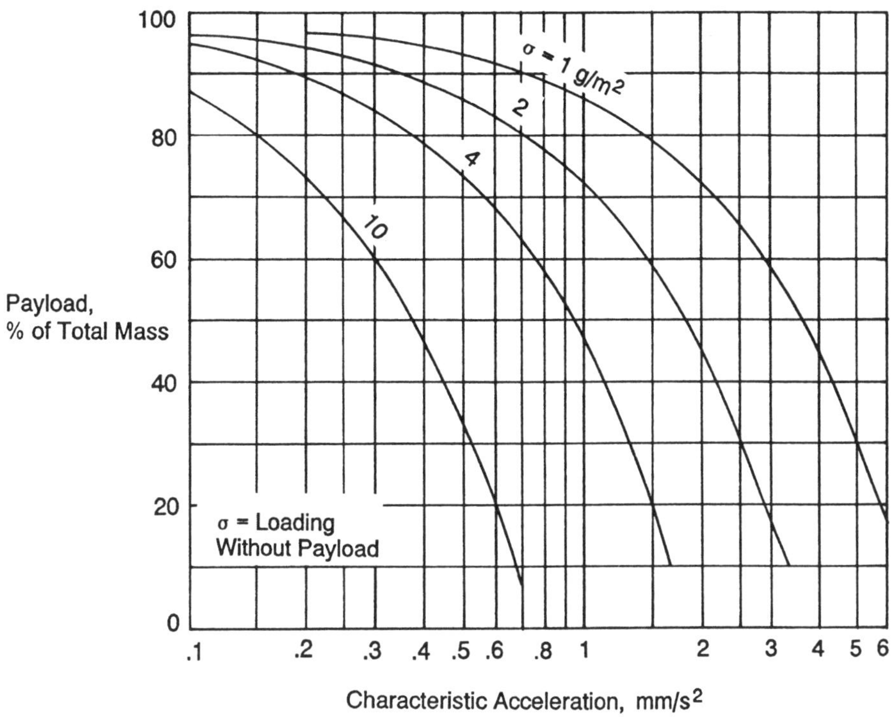

FIGURE 3.2 Payload Variation with Characteristic Acceleration.

thickness and other equal treatment, the square sail has significantly better performance.

It was found with the square sail that very efficient structures could be developed, while the heliogyro was found to need far more load-carrying members than had been anticipated. There is a lesson here which should be heeded in comparing an immature design concept against one which has undergone a substantially more thorough analyses. The mass estimate for an immature design can easily vary by 20% or more during a design development effort, and almost invariably in the direction of more mass. An experienced spacecraft designer could readily consider a budgeted 20% mass margin as highly optimistic for a vehicle as new and radical as the solar sailer. This means that comparisons between design concepts are very uncertain if even one of the concepts is immature. To date, only the square-rigged ship and the heliogyro have undergone substantial design evaluations.

The differences among concepts are most prominent when lightweight payloads are being carried. When payloads are a large fraction of the sailing mass, they predominate in the sail loading so that the advantage of very lightweight sail designs then becomes less important.

Since differences in sail loading can be expressed as differences in transfer times to the destination for a given payload, these differences in sail loading can be expressed as cost differences for specific missions because of the difference in total mission operations costs due to different flight times. The differences in total mission costs are a function of the sail vehicle costs, costs resulting from differences in the mission operation cost rates, and differences resulting from the time of the transfer.

The region of space in which a sail vehicle operates can have a strong influence on its design. This will generally show up in terms of the sail loading and the thermal characteristics of the sail. The design of the first solar sail vehicle will probably allow operation anywhere in the solar system beyond a minimum distance of about 0.3 AU. If a specific vehicle is built to operate only in a restricted region, such as that between Earth and Mars, then that vehicle design can take advantage of the less severe design requirements. This may be most important in the areas of thermal design and accommodating pressure load deflections of the sail and supporting structure. It seems certain that a single design could be capable of operating in the region between Mercury and the outer planets.

Sails and ships can be designed to operate quite close to the Sun, probably significantly inside of 0.2 AU from the Sun. However, designs that can accommodate the resulting load and temperature requirements may not be suitable or desirable for applications in the outer regions of the solar system. Sails made with thin-film technologies for high reflectivity and emissivity may be able to operate in the regions inside of 0.1 AU.

The operations necessary for a ship will impose some control and maneuvering requirements. In particular, these can include operations in planetary orbits, docking, rendezvous, and stationkeeping. These can be strong selection factors.

The complexity of construction, deployment, and support facilities can be important cost drivers that need to be included in a selection assessment.

DESIGN PRINCIPLES

Several general design principles apply to all candidate sailing ship configurations. Those peculiar to specific configurations are treated in subsequent sections. The ship and sail cannot be treated as fully independent because of their interactions. This chapter describes the ships and some of the interactions with the sail, while the next chapter describes more detailed issues of sail design and manufacturing.

Loads on the Sail

The sail requires some means of keeping it spread fully open, done by applying tension at various points on the sail. The source of the tension is either a structure that is loaded in compression or masses attached to the sail, creating an assembly which is spun to generate a centrifugal effect. Numerous variants of both basic types have been proposed. The most prominent of these variants are described in the following sections.

A sail can carry only tension loads since the film material has no effective resistance to compression or bending. The supporting structure must provide the stiffness that the sail lacks. The spinning of a sail provides a form of stiffness, which is primarily carried through the tension members in the sail.

If a piece of sail film were spread open in space and released, solar pressure would cause it to accelerate away from the Sun. The sail piece would keep roughly the same shape since all of the sail would tend to experience the same acceleration. If a small mass were attached to the center of the sail, that mass would present a resistance to the acceleration of the sail. Since the sail has no stiffness, it would promptly collapse behind the mass. To prevent the sail from collapsing, some method of providing rigidity must be used. Additionally, to provide minimal sail deflection with minimal weight of the supporting structure, the load from the attached mass should be distributed evenly across the sail.

It is not practical to have a sail that does not deflect, nor to have a design that can distribute loads completely evenly across the sail. The supporting structure of the sail will always undergo some deflection, and the sail will then experience both large scale and local deflections which result in billowing and wrinkling, respectively.

The problem of the attached mass is, of course, a real one since a sailing ship is intended to carry a payload that typically will be a very concentrated mass compared to the size of the sail. If a ship uses a compression-loaded structure, a straightforward approach is to attach the payload and the operations module to the supporting structure of the sail, Figure 3.3. The loads of the attached masses as well as the loads from the support structure are carried to the sail through the tension members which attach the sail to the structure. The greater the number of attachment points, the more uniform the load distribution will be, reducing the extent of billowing and wrinkling. However, the number of structural members and other factors limit the number of attach points for the sail.

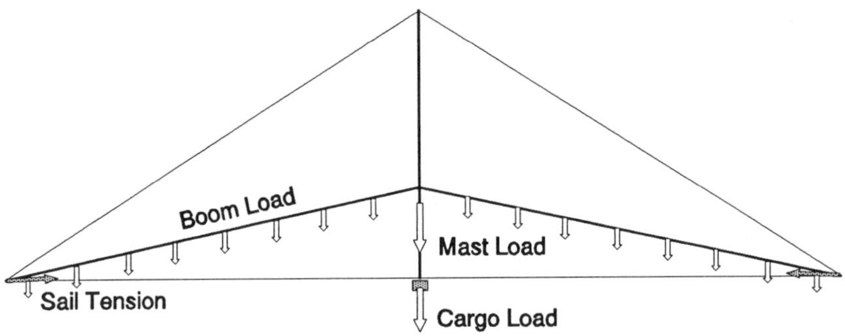

FIGURE 3.3 Load Placement on Ship's Structure.

Structural Loads

Determination of the loads transferred between the sail and supporting structure is necessary to determine the deflections of the sail and the structure. The actual stress loads in the tension members between the sail and supporting structure are functions of the tautness of the sail as well as the net force resulting from solar pressure. The actual loads transmitted to the structure are greater than the transmitted part of the solar force because of the tension loads required to hold the sail taut.

It is not correct to simply take the total pressure across the sail, convert it to a force, and say that this is the load which must be supported by the structure. This approach would result in overstating the actual loads in the structure and sail because of an imbedded assumption that all of the masses, including the structure, are at rest inertially.

Following the principle of D'Alembert, all of the masses except the sail itself are multiplied by the ship's acceleration to obtain the fictitious, or virtual, force of D'Alembert, which represents the inertial resistance of the non-sail masses to acceleration. This D'Alembertian force is delivered to the sail at one or more points as shown in Figure 3.3. This is the acceleration load that is transferred between the sail and the supporting structure. Knowing this, the correct structural loads and sail deflections can be determined.

Another way to determine the correct acceleration load transferred between the sail and the supporting structure is to take the total pressure across the sail, convert it to a force, and then subtract the product of acceleration and sail mass. This yields the same value as the product of acceleration and all masses except the sail. In this approach, we can think of the force supported by the structure as being the total applied solar force on the sail less that part of the force absorbed by the inertia of the sail. This can be seen in the D'Alembertian expression

$$F_{sail} - m_{sail} \cdot a - m_{loads} \cdot a = 0 \tag{3.3}$$

Additional loads in the structure result from the mass of the structure. Since the vehicle is under acceleration, that mass acts like weight in a gravitational field. The analysis can be done considering the structure to be carrying the total load transferred from the sail while in a gravitational field having the same acceleration value as the ship.

If the structure and payload have no mass, the loads on the structure are only those from the simple tautness of the sail, because the inertia of the sail absorbs all of the solar force. If the ship is restrained so that it does not accelerate, then the structure experiences the full effect of the solar force combined with the sail tautness.

Sail Curvature

Ideally the sail would be flat, but in actuality it develops curvature under pressure. Curvature reduces the pressure on the sail, causes a redistribution of the pressure, and can result in local temperatures increasing from concentration of solar energy in the troughs of wrinkles.

The sail is kept taut by the tension applied at various points on the sail. High tension forces tend to keep the sail flat, but result in high stress levels in the sail and large loads in the structural members supporting the sail. Large stress levels in the sail can result in wrinkles. Large loads in the supporting structure result in increased structural weight in the design.

Stress levels in the sail vary with distance from the Sun and angle from the Sun. An increased pressure on the sail increases the amount of stress in the sail and increases the load on the supporting structure. The supporting structure may undergo greater deflection under the increased load, which will increase the curvature in the sail.

If the large-scale curvature in a sail increases during a mission, localized wrinkles will develop. They can cause hot spots and possibly affect the distribution of pressure across the sail. The change in the large-scale curvature is a concern primarily because of the resulting large-scale redistribution of pressure across the sail. This shift of pressure must be allowed for in the design of the control system since it alters the balance of the ship. Both the billow (or cone) and the wrinkles act to reduce the total pressure compared to a less curved sail.

Pressure and Force

The total force and the distribution of force across the sail must be analyzed to accurately determine the accelerating force on the ship and the moments, or torques, that the control system must accommodate. The analysis is based on a numerical integration or summation across the sail. To do this, the sail is considered to be made up of many small facets, each of which is treated as a local flat reflector. The orientation of each facet with respect to the Sun is determined by combining the orientation of the ship relative to the Sun with the orientation of the facets relative to the ship. This latter part is done by modeling the curvature of the sail in sufficient detail to define the orientation of each facet.

The pressure on each facet is a function of the reflectivity, the thermal emissions, and the angle of the Sun from the perpendicular of the facet. The details of this function are given in Appendix A. The product of the pressure and the area of the facet gives the force on the facet. The forces on the individual facets are combined to yield the total force and the force distribution on the sail. This process is described in Appendix B.

Control and Balance

The total force vector, including all sails, vanes, and other significant reflecting surfaces, must pass through the ship's center of mass and point in the desired direction to sail under steady conditions. The curvatures in all sails and all deflections of the structure must be taken into account in determining the total force vector.

The ship's pitch and yaw angles are changed by causing a temporary misalignment of the total force vector and the center of mass. The roll angle is changed by setting sails or vanes so that they create the desired torque about the roll axis.

THE SQUARE-RIGGED CLIPPER SHIP

The square-rigged ship was one of the two studied extensively at the Jet Propulsion Laboratory in the mid-1970's. The sail is 820 m on each side, and has a characteristic acceleration of about 1.03 mm/s^2 as sized for the proposed Halley Rendezvous mission. Some more recent design alternatives are included which make this configuration even more attractive. The overall configuration is illustrated in Figure 3.4.

There is no known upper bound on the size of this type of ship. Designers during the early effort thought that about 1 km would be a practical upper limit for a sail that must undergo automated deployment, but a ship using manned assistance could be made to any size. Changes in the operational environment now make it seem unlikely that this type of ship would undergo fully automated deployment.

In this configuration, the sail is supported at the corners by four booms which provide the tension forces necessary to keep the sail open. The configuration is three-axis stabilized, so that its attitude rates are usually kept at or near zero as it moves through space.

The sail can either be packed after fabrication and unfurled when the sail support structure is completed, or it can be fabricated in place to avoid the unfurling requirement.

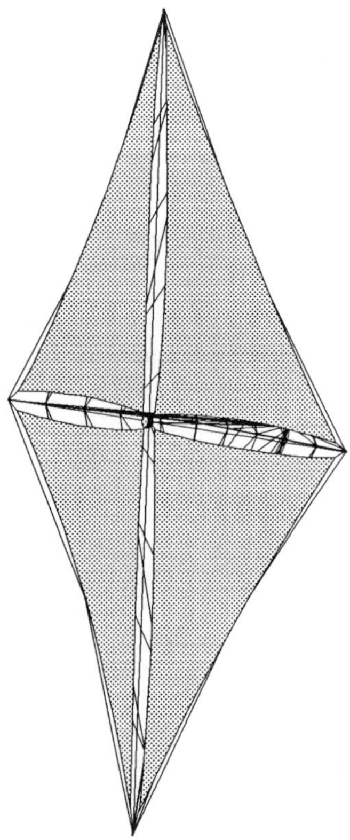

FIGURE 3.4 The Square-Rigged Clipper.

The Clipper Configuration

The configuration presented is a derivative of the design invented at JPL. The major change is in the sail, with the structural concept unchanged. The design has a complex rigging pattern, reminiscent of maritime sailing ships. The rigging provides the benefit of reducing the mass of the mast and booms.

The booms are attached to the mast at less than 90° angles to create the offset structure shown in Figure 3.5. This approach keeps all of the spars and stays on the lee side of the sail, allowing unrestricted deployment of the sail and avoiding double illumination of the rigging by reflected light.

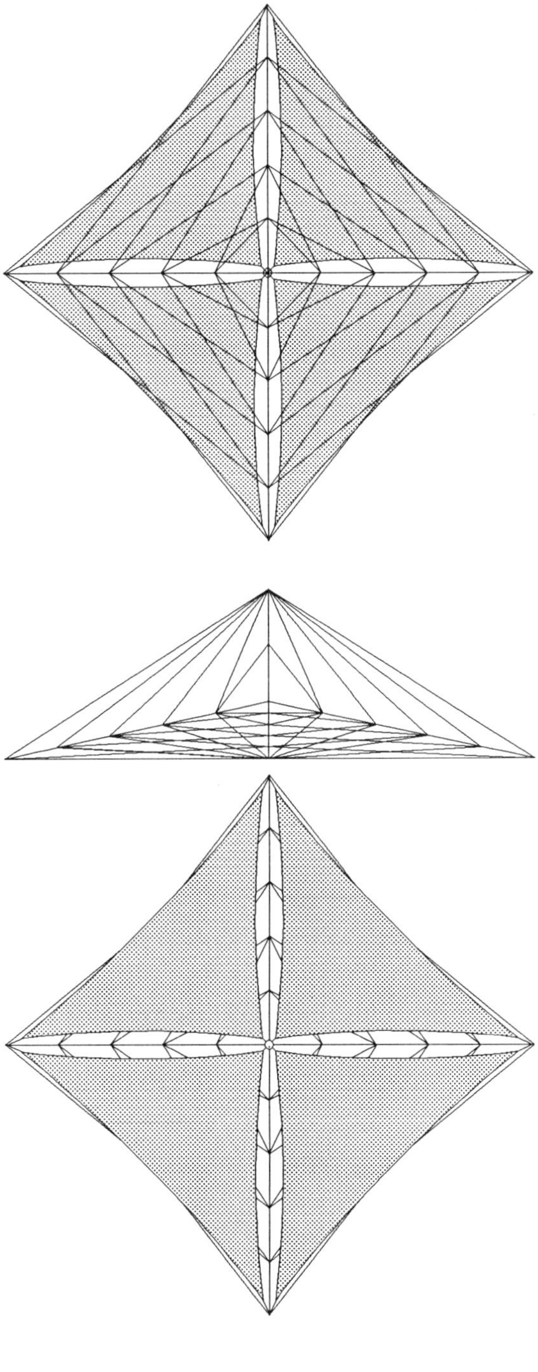

FIGURE 3.5 The Clipper Design.

If a single-piece sail is to be used, a slightly different design must be used to keep the sail from draping across structural elements. In this case, the booms are be pulled back by stays attached to the mast, Figure 3.6. The resulting curvature in the booms helps prevent unpredictable collapse modes in the booms. The payload is attached to the mast near the plane of the sail. This approach was taken with the Halley Rendezvous square sailer design.

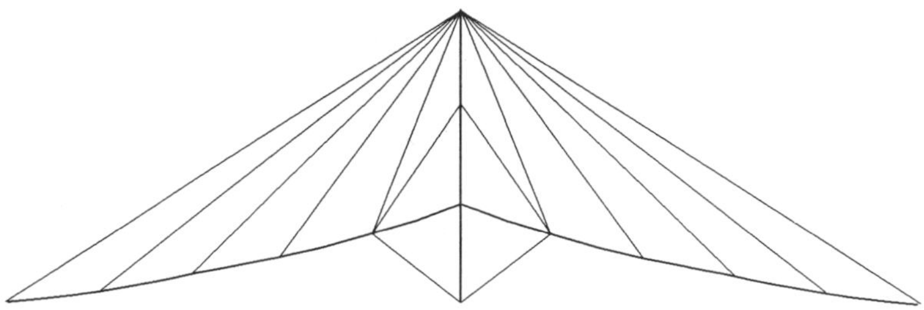

FIGURE 3.6 Clipper with Curved Booms for Single Sail.

The spars are triangular lattice structures made of metal or composite material that can be collapsed for storage in canisters. In the Halley Rendezvous design, the spars have a diameter of 1.2 m (46 in), and are made of titanium to withstand the thermal environment near the Sun.

The stays are lines running between pairs of booms and between the booms and the mast. They stabilize the booms against buckling under compression loads caused by tension in the sail. The Halley Rendezvous design has stainless steel tapes for the stays, but other materials can be used, especially in less severe thermal environments. Alternate rigging patterns are possible, and were considered for the Halley Rendezvous design. The shoelace pattern was selected because of its suitability for automated deployment.

Other square sail configurations are possible. None of the alternates identified have been found to have mass properties better than the reference design.

Sail Shape

The best plan form for the square-rigged clipper appears to be that of four triangular sails spread between the booms, rather than a single full-span sail,

Figure 3.7. The advantages of the smaller sails are better ship control, less wrinkling under load, and easier deployment, replacement, and manufacturing.

The large single sail is a workable approach. The Halley Rendezvous design is based on a single sail, and was subjected to analyses the verified its suitability.

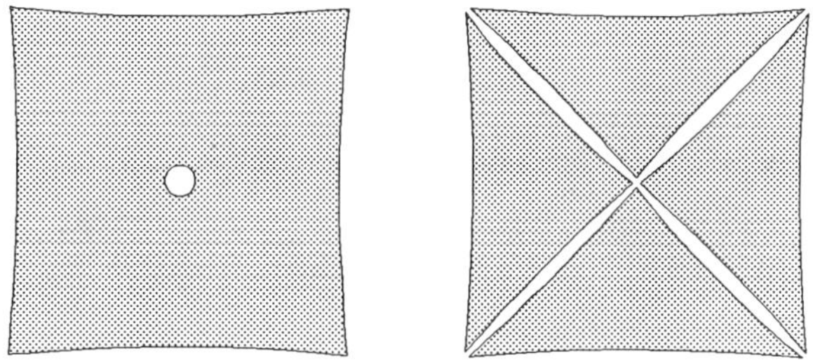

FIGURE 3.7 Sail Patterns for the Square-Rigged Clipper.

Mass Estimates

The mass estimates provided here are for three versions of the square-rigged clipper configuration. The first estimate given in Table 3.1 is the Halley Rendezvous 820 m sail design, based on a 2 μm sail. The second is a quarter-sail configuration, with the sail film thickness reduced by 50% and related adjustments made to other parameters. This may be representative of a ship having its sail fabricated on Earth and deployed at a spacedock in low Earth orbit rather than by automated deployment. The third, probably optimistic, estimate is the same quarter-sail configuration except that the sail has a minimal plastic film. This type of sail would require fabrication in space. The last two cases benefit from not carrying the sail deployment loads or the high loads encountered near the Sun.

The loading values are obtained by dividing the mass estimates by the actual sail areas of the sails (641,200 m^2 for the Halley Rendezvous sail). The total sail loading in the cruise configuration must also include the averaged loading of special hardware and the payload. The figure shows the loading value of the sail system, excluding the loading from the operations module, any special equipment, and the payload.

TABLE 3.1 Mass Estimates for 820-Meter Square-Rigged Ships.

DESIGN:	*HR-820*	*Clipper*	*Ultralight*
Sail film[a]	1821	824	162
Reflective layer[a]	173	157	157
Emissive coating[a]	58	52	52
Sail tendons	38	35	35
Mast and booms	760	400	80
Boom support stays	130	80	20
Stay reels and tensioners	50	40	30
Boom positioning hardware	80	70	50
Sail form control mechanisms	50	50	40
Contingency (20%)	222	135	51
TOTAL MASS, kg [b]	3382	1843	677
Area of sail, m^2	641,200	580,000	580,000
Sail loading, σ, g/m^2 [b]	5.27	3.18	1.17
a_c upper limit, mm/s^2 [b]	1.54	2.55	6.94

[a]excluded from contingency [b]excluding operations module and payload

THE QUADSAILER

This configuration uses four fully separate sails, each supported by a pair of articulated spars, Figure 3.8. Through movement of the spars, each sail can be set to any fore-and-aft position and turned around its axis. This design achieves maximum controllability, but at the expense of substantial mass in the structure required for the sail. Pitch and yaw can be achieved by fore-and-aft motion of pairs of sails when at an angle to the Sun, or by bringing a pair of spars closer together to alter the pressure on that sail.

The total length of spars in relation to sail area is greater with this configuration than with the square-rigged ship. Since the spars are not stabilized by stays, they must be more massive to withstand the bending loads from the sail. The result is a sail loading that is significantly greater than the square-rigged ship, resulting in lower performance.

This and the two following configurations may be well suited for intelligent, autonomous explorers, to be set loose to roam through the solar system.

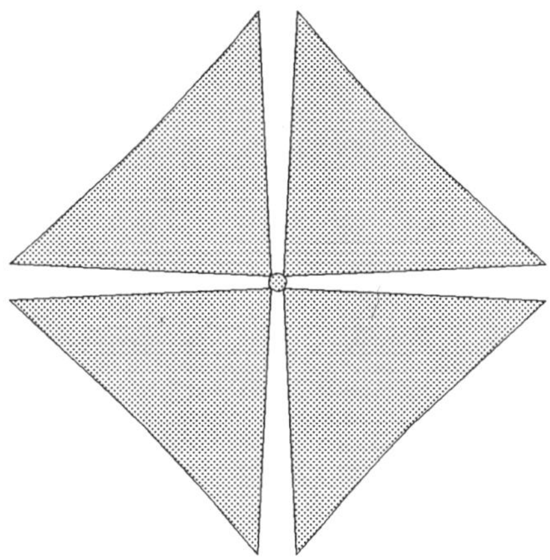

FIGURE 3.8 Quadsailer Configuration.

THE TRISAILER

The trisailer uses three separate sails similar to those shown for the quadsailer, and is fully controllable in all axes.

Two configurations are shown in Figure 3.9. The second type uses a central spar to gain more sail area and to possibly serve as a hinge line. The resulting gain in sail area is small compared to the mass of the central spar, so this design has a lower performance than the quadsailer. The additional control mode from having the central spar does not add any essential control capability since the ship is fully controllable with the three simple sails. The second configuration could be controlled by only fore and aft movement of the edge spars of each panel.

THE BUTTERFLY SAILER

This configuration has two sails somewhat like butterfly wings. A center spar is necessary in each sail, Figure 3.10. The edge spars can provide full control by lateral or fore-and-aft movement, either along the entire lengths or only the outer parts of the spars.

Instead of edge spars, somewhat shorter spars could be placed

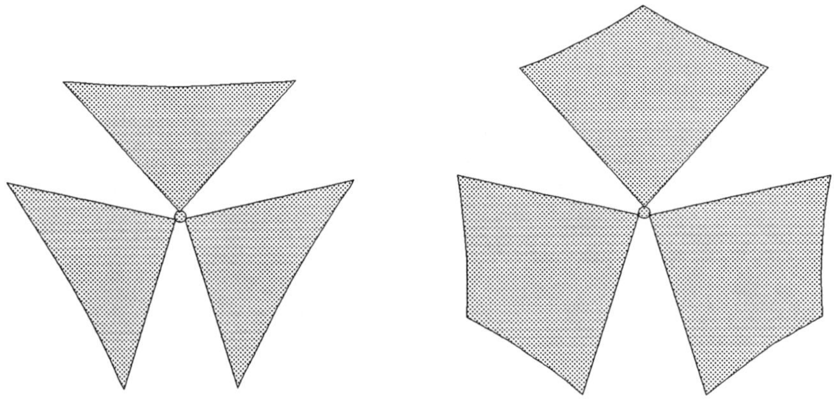

FIGURE 3.9 Trisailer Configurations.

perpendicular to the center spar, giving essentially the same sail shape. This later concept might reduce the bending loads on the lateral spars, with a resulting weight savings, but probably then necessitating use of tip vanes. Use of crosstrees near the spar tips allows using the tips as simple vanes to provide attitude control.

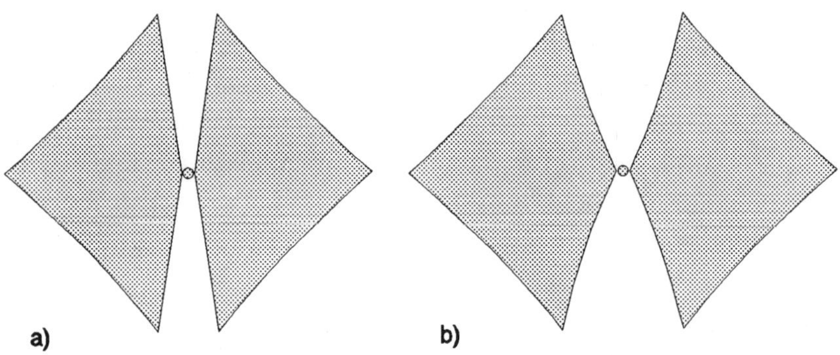

FIGURE 3.10 Butterfly Sailer Configurations: a) with spars at inside edges, b) with perpendicular spars.

THE RIBBED SAILER

This configuration, considered briefly during the JPL effort, relies on flexible ribs that wrap around a central hub, Figure 3.11. This configuration is based on several antenna designs that have successfully operated in space. The strain energy in the ribs provides the deployment forces. As the ribs unwrap, the sail, or set of sails, can be pulled open simultaneously. The hub needs to have a diameter that is a substantial fraction of the overall dimension of the sail, with an attendant mass penalty unless a means is provided to jettison the hub after sail deployment. The ribs will be massive compared to the booms in the clipper. If a set of individual sails is used as illustrated, the sails can be used to provide attitude control without the use of vanes.

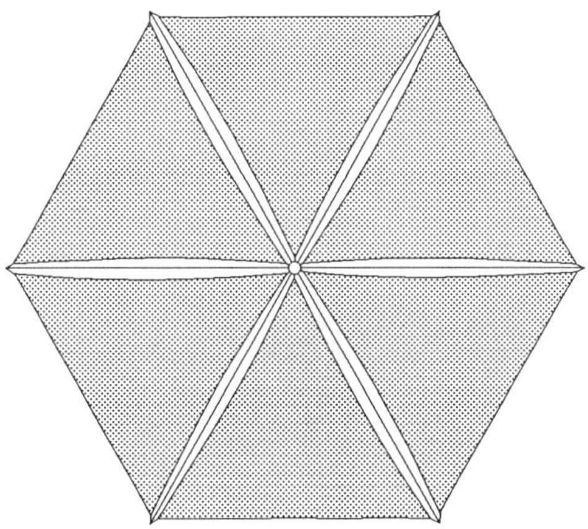

FIGURE 3.11 Ribbed Sailer.

THE RING SAILER

The sail has a circular shape, supported around its perimeter by a ring which provides the means of keeping the sail open. A simple parachute is not usable because the shroud lines will cause it to collapse. Solar pressure is not an analogue of aerodynamic pressure.

In one configuration of ring sailer, the operations module and payload are attached to the ring by long shrouds. An additional shroud attaches to the center of the sail to reduce the billow. Without this line, the sail would billow out into a simple concave shape which could reduce the controllability of the ship. Additional attach lines to the sail might be helpful in improving the efficiency of the configuration.

The stability and control of the configuration does not appear to be amenable to a simple analysis. A definitive analysis must include the force and moment of the sail, the distance to the operations module, and the method of attachment of the operations module to the shrouds. To date, no such analysis has been done, leaving the practicality of controlling the configuration an open issue. Preliminary analyses done at JPL indicated that the configuration might be difficult to control.

The second configuration uses a central mast with the midpoint of the mast at the center of the sail, shown in Figure 3.12. Stays run from each masthead to the rim to provide definite control of the sail and to stiffen the ring.

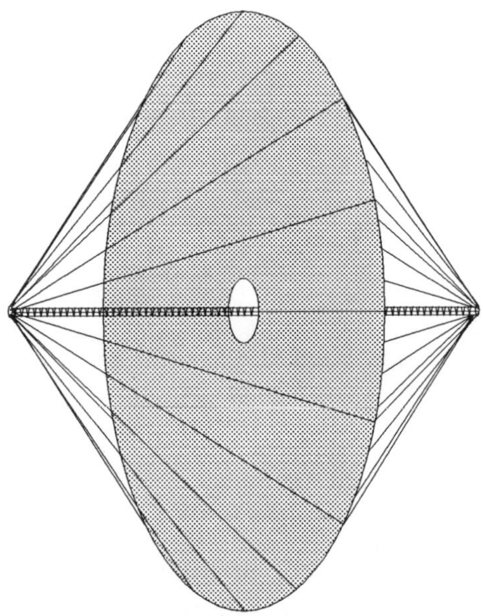

FIGURE 3.12 Masted Ring Sailer.

Two types of rim structures have been considered for this configuration. A thin ring is subject to warping if not constrained, so the rim must either be held by the sets of fore and aft stays shown in the figure, or it must be built with depth provided by additional structural members. Both of these approaches result in relatively inefficient structures. One type of rim is a lattice beam structure, which was found to result in a greater structural mass than that of the square-rigged ship.

The other type of rim structure is an inflated tube, which could result in a low structural mass. However, inflated structures are unacceptable for sustained operations because of the inevitability of punctures by micrometeorites.

As a result of the relatively poor mass properties, the ring-supported configurations were not subjected to any in-depth analysis.

THE DISK SAILER

This concept of a circular sail uses the centrifugal effect from spinning to maintain sail shape. This configuration received a modest investigation during the JPL study, which defined a 860 m diameter sail as a candidate for the Halley Rendezvous mission. This concept was meant for assembly on Earth with automated deployment in space. A design meant for assembly or fabrication in space could have some substantial differences from this design. Many variants are possible with the basic design presented here.

The disk sailer has the potential to provide good performance, depending partly on design cleverness and partly on sail materials technology. The structural mass of spinning sails is partly a function of the sail mass, so that a thinner sail could provide some benefit in structural mass. However, use of thinner sail film could require the sail being made in smaller panels, increasing the complexity of the sail.

Configuration

The basic configuration of the sail is that of a circular disk, although the perimeter could have straight segments, or some inward curvature, as shown in Figure 3.13. A set of sails should be used rather than a single, solid sail, for several reasons, including deployment handling, control, replacement, and sail smoothness during coning.

The central portion of the disk is open, with a diameter about 10% of the overall diameter, shown in Figure 3.14. A short central mast, or spindle, anchors tension lines serving as spokes that stabilize the hub rim and transfer precessional torques between the center body and the sail. The spindle length should be at least about 15% of the central opening. The lines from the inner rim to the spindle attach to disks at each end of the spindle. The lever arms provided by these disks transfer spin torques between the sail and the spindle. The diameter of these disks might be from 5 to 10% of the central opening.

The segments of the inner rim can carry the furled sails. The segments could be long rollers that provide a controlled deployment of the sail, or the sails can be folded and constrained in some manner by mechanisms on the segments.

A multi-mission sailer will need a hub that allows docking on both sides of the sail, provides a connecting tunnel between the docking hatches, and permits internal maintenance to be done in a pressurized environment. A payload bay can be provided on the sunward side, primarily for thermal protection of the payload. The central tunnel can be despun, easing the task of docking by other spacecraft.

Pressurized compartments can be placed radially outward from the hub to provide artificial gravity environments if people are carried on the ships for long voyages.

Space Liners and Sailing Cities

The illustrations in Figures 3.13 and 3.14 purposefully have no dimensions shown because the designs can be scaled across an immense design range. The disk sailer might be the best configuration for scaling up beyond the kilometer-sized ships we currently think of designing.

Consider a 1 km toroidal structure having a mass allowance of 2000 tons. This allowance is 637 kg per linear meter of the circumference (428 lb/ft). Using steel or aluminum walls wrapped with high-strength fibers, the diameter of the pressurized shell could be 5.6 m (18 ft), as an example. The mass allowance for everything inside is then about 700 to 1000 tons. The shell is large enough to allow two floor levels with suites of offices and apartments, plus manufacturing and agricultural space.

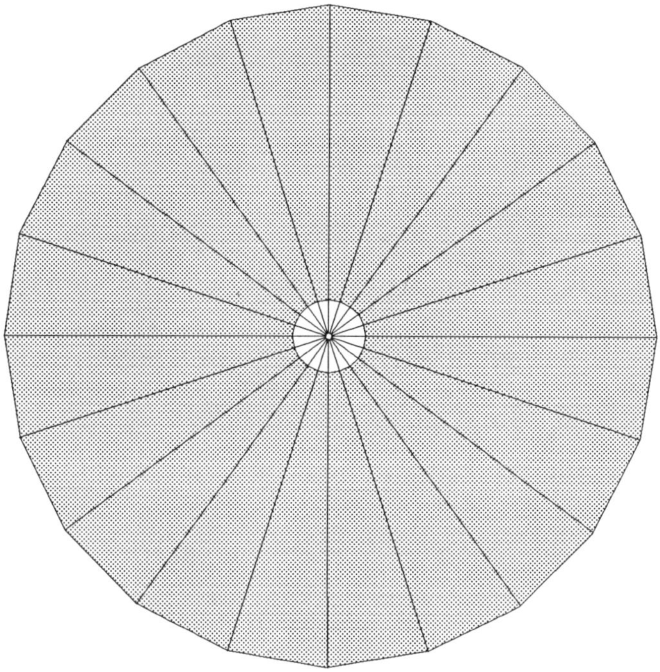

FIGURE 3.13 Disk Sailer Configuration.

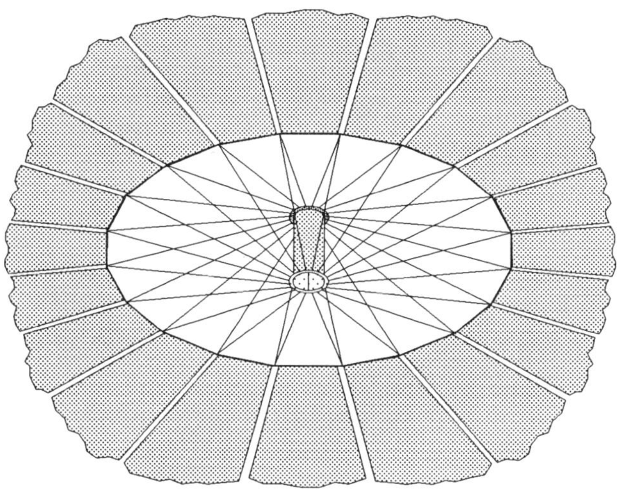

FIGURE 3.14 Disk Sailer Central Region, perspective view.

Spinning the structure with a period of 72 seconds provides an artificial gravity equal to that of Mars (38% of Earth's gravity).

This structure serves as the inner rim of a disk sailer. With sails extending out to 10 km, this space liner could cruise with a characteristic acceleration of about 0.37 mm/s^2. With larger sails extending to 14 km, the characteristic acceleration is about 0.68 mm/s^2.

Keeping the same acceleration levels with larger structures, a 2 km toroid would have a mass allowance of 8000 tons, or 1270 kg/m (856 lb/ft). A 4 km toroid would have an allowance of 32,000 tons, or 2550 kg/m (1710 lb/ft).

Growing the dimensions in this manner can take us from the realm of large ships to the realm of mobile cities. We can start with the concept of a toroidal space city and turn it into a sailing city at any time by the addition of sails to the outside of the torus.

Shape

The pressure exerted on the surface by sunlight combined with the concentrated mass at the center of a disk sail causes the sail to form into a slightly conical shape. This coning becomes more pronounced closer to the Sun, and may induce substantial wrinkles in the sail. At a distance of about 0.25 AU from the Sun, the Halley Rendezvous design has an 8 m deflection at the rim, about 1% of the diameter.

To avoid problems associated with a solid sail, the sail is divided into panels linked periodically along their edges, Figure 3.15. The links are massy chords that flex outward under the centrifugal effect to draw the panels tight. Designs might conveniently be based on using from 6 to 24 panels.

Mass

Mass estimates using JPL data are given in Table 3.2, which shows both a nominal version, and a more optimistic version. These cases are based on using film thicknesses of 2000 and 1000 nm, respectively. Other mass values are kept the same in the comparison.

The values in the table indicate better performance than with the square sail or the heliogyro, which is expected because the disk sailer uses less structure than the other two types. The disk sail estimates are less mature than the others, so there is greater uncertainty in these mass estimates.

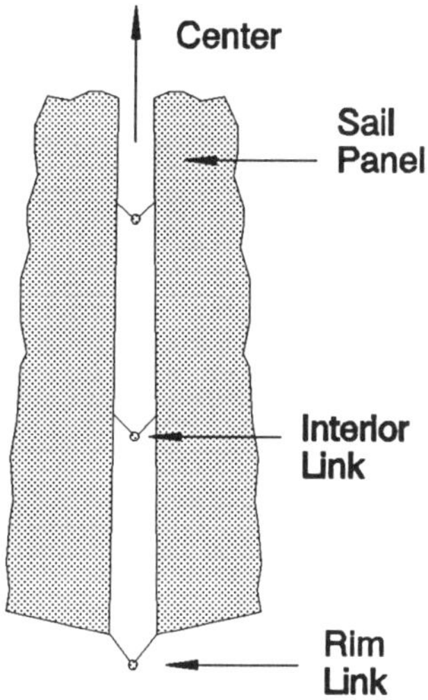

FIGURE 3.15 Sail Panel Links.

TABLE 3.2 Mass Estimates for 860-Meter Disk Sailers.

SAIL FILM, nm	2000	1000
Sail[a]	1880	1061
Edge tendons	64	64
Joint ties	106	106
Rim structure	50	50
Spokes	104	104
Spoke reels	16	16
Truss	7	7
Deployment mechanisms	97	97
Contingency (20%)	89	89
TOTAL MASS, kg[b]	2413	1594
Area of sail, m^2	576,600	576,600
Sail loading, σ[b]	4.18	2.76
a_c, mm/s^2[b]	1.94	2.93

[a]excluded from contingency
[b]excluding operations module and payload

THE LATTICE SHIP

This type of ship is among the most complex to construct and operate, but it may also be the one of the best performing designs. This design concept has not been subjected to a thorough, detailed analysis, leaving the estimates of its characteristics more uncertain than those of the clipper, disk, and heliogyro sailers. The ship, illustrated in Figure 3.16, was invented by Eric Drexler.[1]

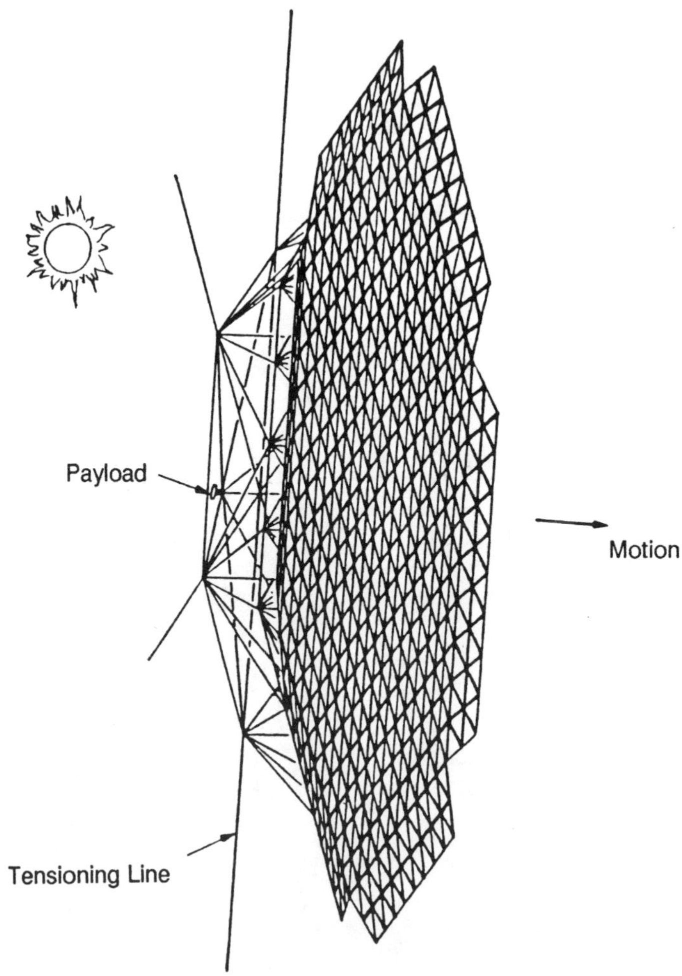

FIGURE 3.16 The Lattice Ship. (K.E. Drexler)

An advantage of the lattice structure is a more even loading distribution from the payload and operations module. The sail is held flat on a large scale without the coning inherent in the disk sailer. However, since the simpler disk sailer typically has only a small deflection, it might be difficult to justify the construction and operations complexity associated with the lattice sail. A drawback of the design is that the lattice rigging, the operations module, and the payload are subject to double illumination since they trail behind the reflective sail, which could cause some thermal problems.

The lattice sail will require constant acceleration to keep the lattice structure under tension, which could be a substantial problem for interplanetary ships. It may be impractical to secure a lattice ship to a dock for the transfer of people and cargo.

Configuration

The lattice ship can be thought of as a variant of the disk sailer, although part of its structural concept is substantially different. The basic pattern of the sail is similar to the disk sail, as shown in Figure 3.17. The primary difference from the previously described disk sail is the existence of numerous attach points distributed across or around the sail, and absence of the central disk structure.

The lattice structure is a network of tension members connecting the operations module and payload to the sail, illustrated in Figure 3.18. The acceleration loads from the operations module and payload are distributed to many points across the sail, keeping the sail relatively flat.

The sail and structural network rotate so that both are stiffened by the centrifugal effect. All of the structural elements between the operations module and the sail are simple tension lines. Masses are included in the rigging to provide tension in stays running parallel to the sail surface.

The structural network connects to attach points at the outer edge of the sail and attach points across the surface of the sail. The shape of the sail in a frontal view could closely resemble, or even be the same as, a disk sail. The surface of the sail will have a peak at each attachment point of the lattice structure. Between these peaks, the sail will curve in a billowed shape.

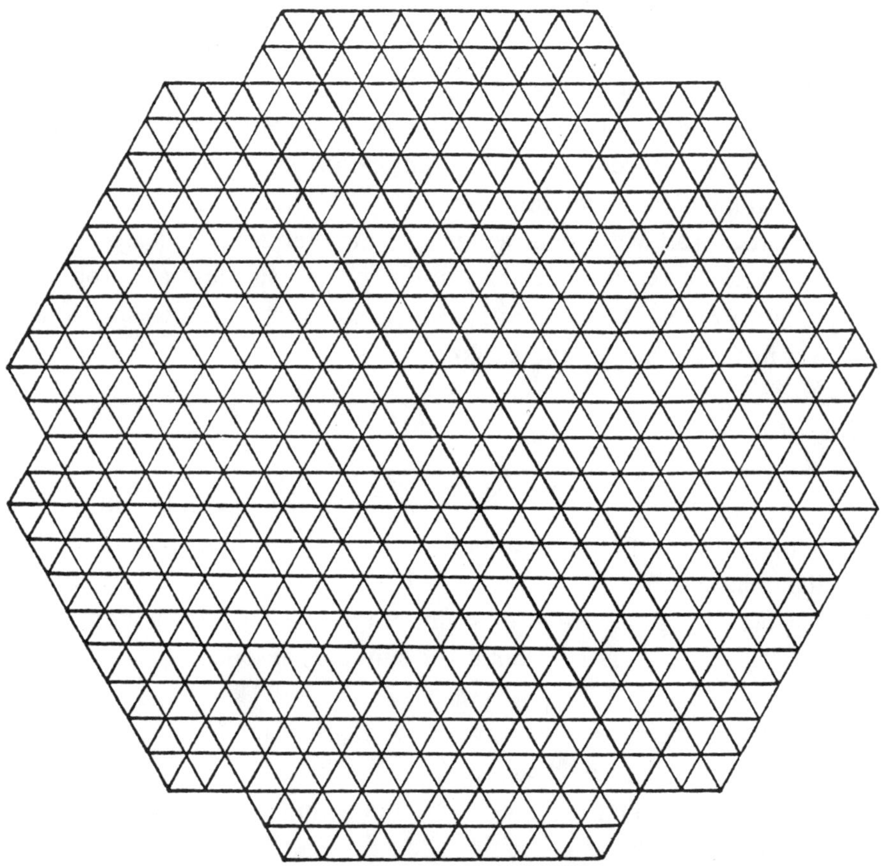

FIGURE 3.17 The Lattice Ship Configuration. (K.E. Drexler)

Mass

An estimate of the mass of the ship is shown in Table 3.3, based on using an ultrathin, pure aluminum sail without a substrate film. Note that no contingency mass allocation has been included. The ship diameter is assumed to be 1000 m, with an area fill ratio of 0.9.

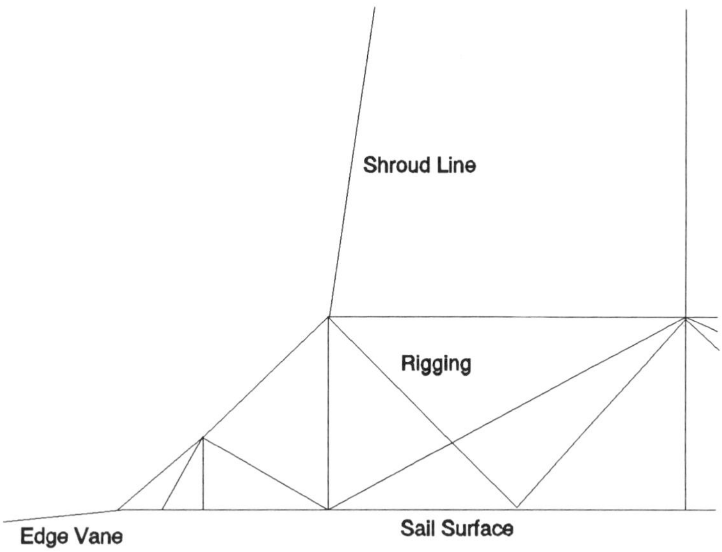

FIGURE 3.18 Lattice Rigging Line Detail.

TABLE 3.3 Mass Estimate for Lattice Ship.

Sail	28
Tendons	21
TOTAL, kg [a]	49
Area, m^2	707,000
σ, g/m^2 [a]	0.07
a_c, mm/s^2 [a]	116

[a]without payload and operations module

THE HELIOGYRO

The heliogyro is a unique sail concept that would operate like a large helicopter rotor spinning its way through space. Its appearance and operational characteristics are much like a helicopter rotor, but it is several hundred times larger.

The heliogyro can be recognized as related to the disk sailer with two differences. The panels in the disk sailer become straight blades that are not interconnected. The heliogyro blades are rotated around their radial axes as the disk panels can be designed to do, but this is the only control technique available for the heliogyro. The result is a sail with some morphological and operational similarity to the disk, but which is truly a unique design concept.

During the design effort at JPL for the Halley Rendezvous sailer, the heliogyro was selected as the baseline design concept on the basis of its deployment simplicity compared to the square-rigged sail. This was in the context of using an automated deployment process, which is a complex operation for the square-rigged clipper.

Configuration

The sail on the heliogyro has a configuration like a helicopter, relying on the centrifugal effect from rotation to maintain the proper shape of the blades. Each blade can be twisted independent of the others to control the attitude of the sailer or to point the force vector.

The Halley Rendezvous configuration has 12 blades in 2 banks of 6 blades, Figure 3.19. The banks are fixed on a single hub so that they rotate together. If the banks counter-rotated, the blades would strike each other because of the large excursions they undergo at their tips. The blades are 8 m wide and 7340 m long, giving the configuration an overall diameter of almost 15 km.

The blades are complex structures in themselves. The centrifugal loads are carried in tension members, called tendons, at both edges of the blades. The arrangement of tendons in the blades is shown in Figure 3.20. The edge tendons are made up of several tapes of high-strength fibers that have periodic crossovers to provide multiple load paths. This gives each edge tendon the ability to withstand several breakages without having a total failure. The edge members also provide the advantage of substantially increasing the torsional stiffness of the blades.

FIGURE 3.19 The Heliogyro. (Clyde Alcott)

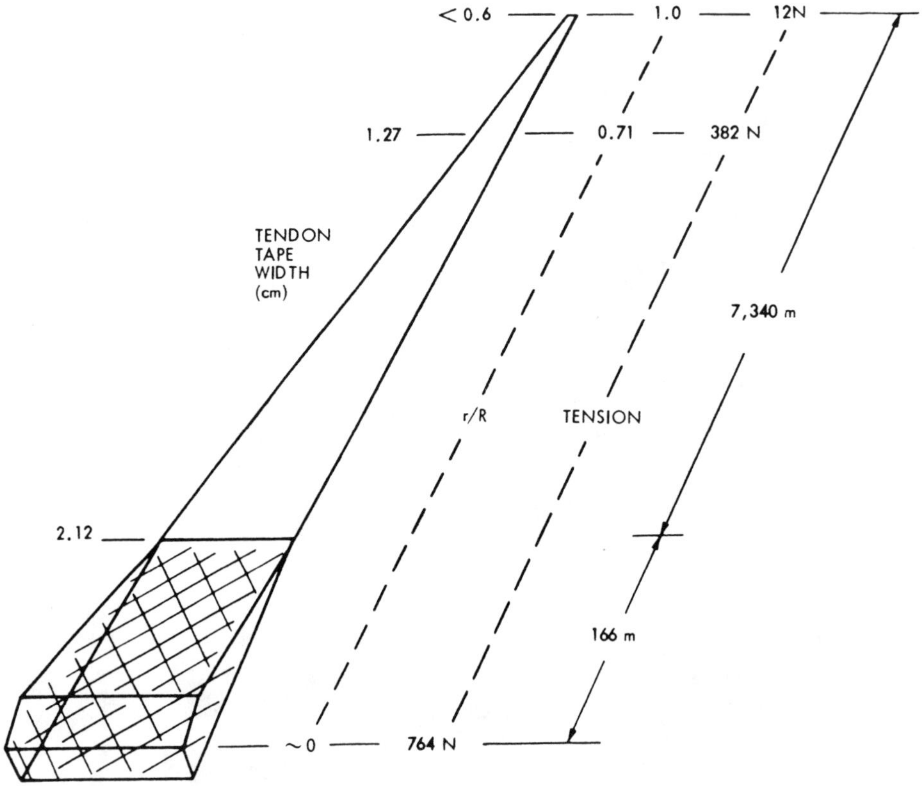

FIGURE 3.20 Heliogyro Blade Tendon Arrangement. (JPL)

The blades begin several meters out from the hub. The inner transverse member is attached by tendons to four arms mounted on the blade roller, as shown in Figure 3.21. This arrangement of arms and tendons aids in controlling the motions of the blades.

Several blade configurations have been evaluated, all using transverse battens to control the edge tendons. Three of the blade configurations are shown in Figure 3.22. Simply attaching sail material between the flexible edge tendons and the battens would lead to a wrinkled sail as the edge tendons pull into catenary shapes. This problem was overcome to a large extent by using small sail panels in the baseline configuration , but at the cost of manufacturing complexity.

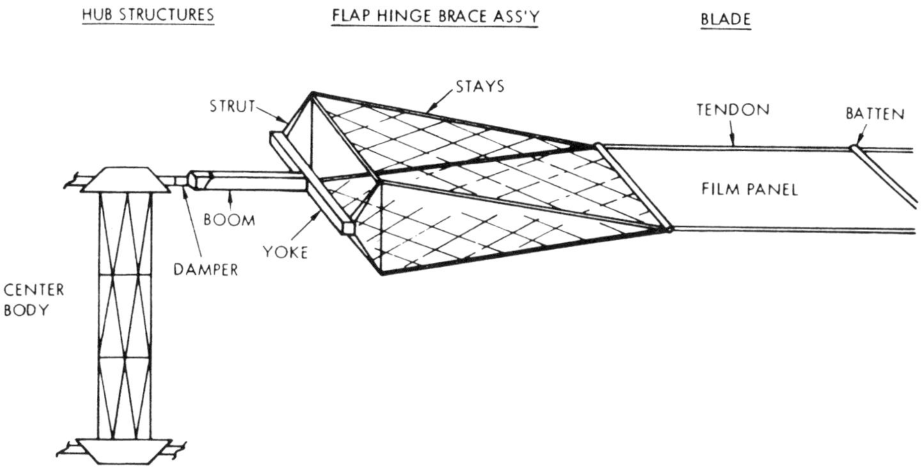

FIGURE 3.21 Heliogyro Blade Attachment Details. (JPL)

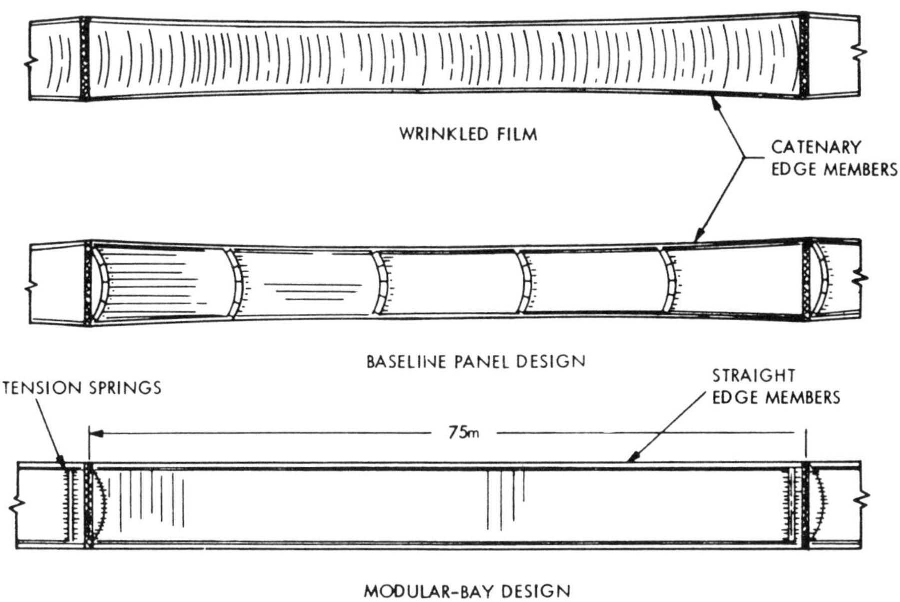

FIGURE 3.22 Heliogyro Blade Designs. (JPL)

Another option is to use edge members that resist bending in the transverse direction to minimize the serious wrinkling associated with large bays. However, this is likely to have a substantial mass penalty.

The transverse battens are compression members used to maintain the correct spacing between the edge tendons. The spacing of the battens varies with position along the blade. The battens are made of high-strength fibers in a lattice configuration. The battens are compressed flat while the blade is rolled up, but spring into shape as they leave the roller. Each batten serves as the upper attach structure for a blade panel.

The reflective panels in the blades of the Halley Rendezvous design are suspended like window shades, as shown in Figure 3.23. The lower (outer) corners are attached to an edge tendon with springs to maintain tension in the panel.

Attitude Control

Attitude control is achieved by twisting the individual blades in a cyclic manner so that the vehicle precesses. A resulting requirement is that all blades be capable of independent twist motion. The control system must be able to coordinate the individual blade motions among the blades and in concert with the spin of the ship. The dynamics and control of the heliogyro are described in Appendix D.

Mass

The mass during cruise is given in Table 3.4 for the heliogyro as designed for the Halley Rendezvous mission. The sail film is 2 μm thick.

INTERSTELLAR SAILING SHIPS

Many methods for reaching Alpha Centauri and other stars have been studied. It still is not clear what approach would be best, but power limitations and propulsion efficacy are two key design factors. The use of beamed power may best solve the power problem, while using the momentum carried in a power beam may best solve the propulsion efficacy problem. Thus the beam-driven lightship is a strong technical candidate for the first starship.

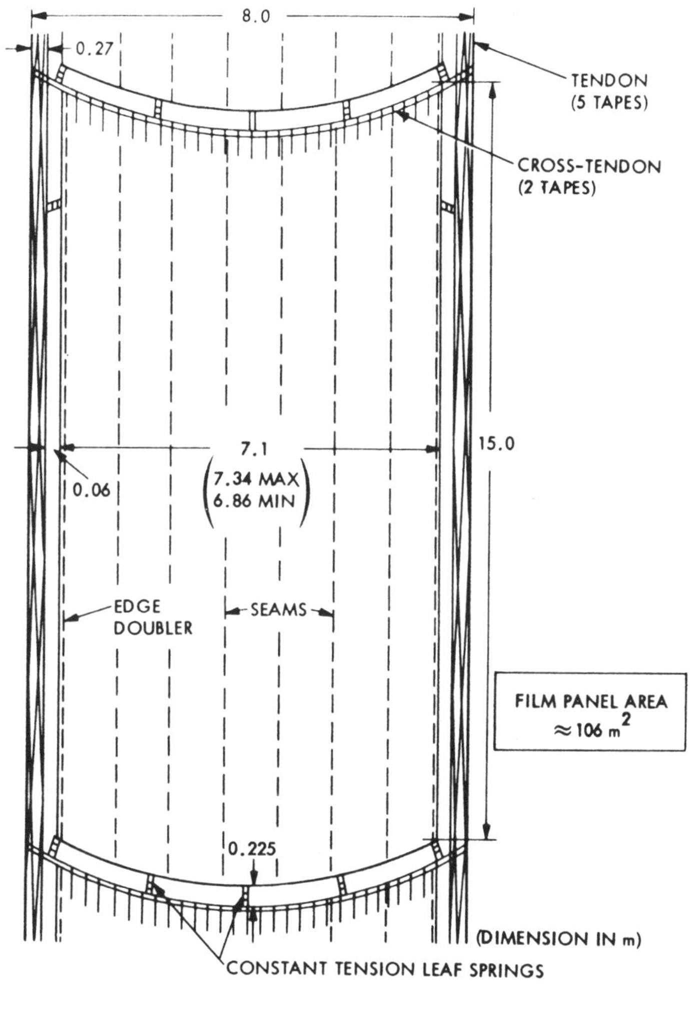

FIGURE 3.23 Heliogyro Blade Panel Detail. (JPL)

TABLE 3.4 Mass Estimates for Halley Rendezvous Heliogyro.

Blade assembly (12)	234.83[a]	2818
Film	147	
Coatings	18.25	
Seams	11.67	
Tendons	42.83	
Tendon coating	0.33	
Battens	7.5	
Tip mass	1.67	
Film tension springs	2.08	
Tuning masses	3.5	
Reel and hinge assembly (12)		195
Boom assembly (12)		102
Center body		635
Miscellaneous		25
Contingency		220
TOTAL MASS, kg [b]		3995

Area, m^2	624,800 (film: 610,968; tendons: 13,832)
σ, g/m^2 [b]	6.394
a_c, m/s^2 [b]	1.27

[a]per blade [b]without payload and operations module

The deceleration phase presents the most difficult problem for a sailing ship. The most likely appearing solutions are: 1) use of a simple configuration traveling sufficiently slow that it can use the light pressure from the dual suns of Alpha Centauri combined with gravity-assist swing-bys of the suns for orbit insertion, 2) use of a multi-stage sail configuration that can decelerate using the power beam, and 3) use of the beam power to drive engines with high exhaust velocities. The use of beamed power is discussed in Chapter 7.

The Single-Stage Interstellar Ship

A single-stage ship has the least cost and risk, but its travel time is too long to consider without beamed power. Even under optimistic operating assumptions, the ship could require several thousand years to reach Alpha Centauri operating on direct solar illumination.

A single-stage, laser-driven ship is conventional in design, except that it is designed for extreme performance levels. It will be very large if it is meant to carry massive payloads.

The ship cannot stop at its destination relying only on solar pressure when trip times are a few centuries or less. Once an energy beam system is established at the destination for deceleration, the single-stage configuration becomes the design of choice.

The Hybrid Interstellar Ship

Propulsion systems based on mass expulsion grow exponentially in initial mass as the velocity requirements are increased. This is a major problem for an interstellar ship that must gain sufficient speed to have an acceptable trip time, and also stop at its destination. Carrying propellant only for the deceleration allows a major reduction in initial ship size.

This approach can be implemented by using laser-driven sails for the first phase of the voyage, so that no propellant is required for gaining speed. Carrying the propellant during the acceleration phase reduces the acceleration of the ship, or requires a greater beam power than is otherwise needed. However, these impacts may be small enough to make this a viable approach. It might also be possible to accelerate without the propellant mass by relying on collecting the deceleration propellant during the cruise phase of the voyage.

The Multi-Stage Interstellar Ship

It may be practical to make direct use of the light pressure in the laser beam for decelerating through a back reflector to reverse the incident light. This can be achieved by using a two-stage sail, where the outer part of the sail detaches from the ship to become the back reflector. The reversed light beam then impinges on the sail of the decelerating the ship.

Two and three stage sailing ships have been proposed as a means of accomplishing interstellar travel.[2] The concepts are based on the disk sail as the basic configuration, with the smaller stages attached at the center of the largest, as shown in Figure 3.24.

The cost, complexity, and risk of the multi-stage sail are difficult factors to evaluate, but the performance advantage of the design may make this the

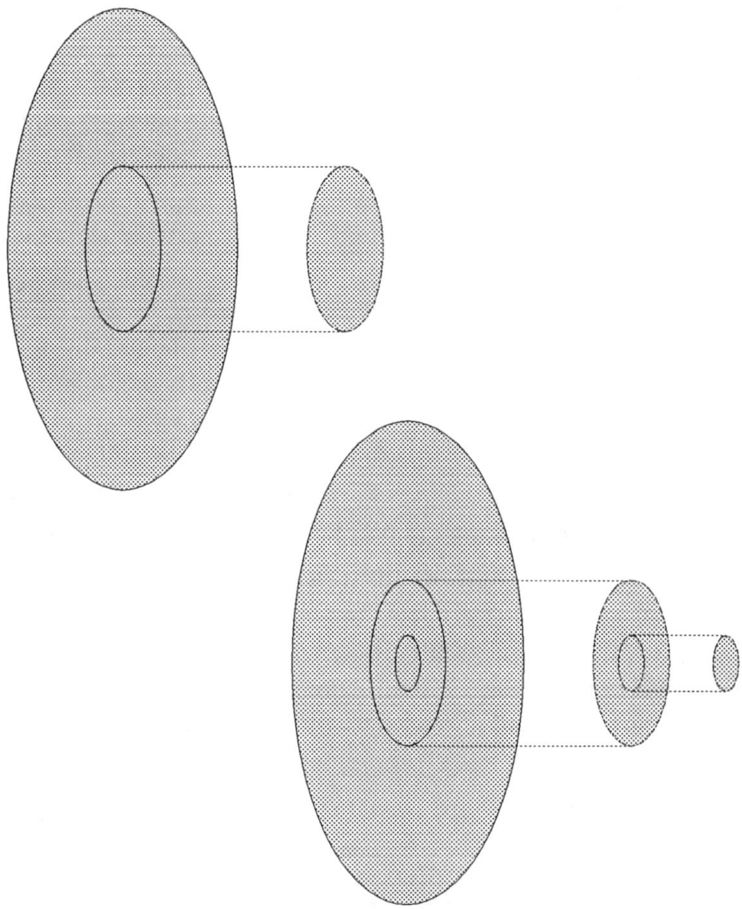

FIGURE 3.24 Two and Three Stage Interstellar Sailing Ships.

configuration of choice. The complexity of the ship may make it desirable to have a crew onboard, but the risk may make it desirable to have the first few ships roboticly operated. Resolution of this uncertainty will depend on the robotics technology existing at the time of construction.

The required shape of the sail is a function of the mission phase. Initially, all surfaces are to be flat or nearly so. After a sail separation, the larger sail must be configured as a focusing reflector. This presents a design challenge, but it appears to have at least conceptual solutions.

Control of the ship and the reflector when it is in use are critical design factors because of the relatively narrow light beam that the ship must follow. This is discussed further in Chapter 7.

The ship and the reflector must have control mechanisms that are robust and fast acting. At least some of the control techniques described for the disk sailer appear likely to provide adequate solutions. The control system must accommodate the nature of the curved reflector not having an inherently stable shape. If an angular displacement develops between the reflector and the light source, the reflector has a tendency to continue turning away from the beam.

COMPARISON OF DESIGNS

One of the major discriminating factors between configurations is the achievable sail loading before the addition of payload. The lack of existing substantial design work on some configurations prevents making a reliable comparison between all of the various types of sailing ships at present.

The use of tension structures, as with the spinning sails, is often more efficient than the use of compression structures. However, the heliogyro was found to be less structurally efficient than the square-rigged ship. It might be that some variant of the disk sail will be found to have the best advantage in sail loading, but it will take substantial analysis, and perhaps even operating experience, to make this clear.

The square-rigged clipper ship has definite advantages in operational simplicity and in maneuverability. Constructing the ship at an orbital facility assures a reliable product. This appears to be the best type of ship for maneuvering near facilities in space, or for stationkeeping with small bodies in solar space if required.

As long as the mass of the sail film dominates the mass of the ship, differences among configurations will likely remain minor with regard to performance. In the case of very thin or nonexistent sail film, the various configurations could have important differences in performance. Among the configurations, there are many uncertainties in implementation which can only be resolved by testing in space.

Cost comparisons should include the direct ship cost, operating cost, and specialized infrastructure cost. Unfortunately there is no cost data to allow comparisons among the various configurations. The square-rigged clipper appears to be a good candidate since it requires no external frame for its deployment, and it can operate from low orbit with a tether, as discussed in Chapter 6. The heliogyro might also be a good low-cost candidate. The lattice sailer might have a low unit cost, but it would require the use of a full diameter construction frame in high orbit. This might make it an unattractive candidate until the space infrastructure is built up.

REFERENCES AND NOTES

1. Drexler, K.E. (1977) Design of a High Performance Solar Sail System. *MS Thesis*, Dept. of Aeronautics and Astronautics, Massachusetts Institute of Technology, Boston.
2. Forward, R.L. (1984) Roundtrip Interstellar Travel Using Laser-Pushed Lightsails. *Journal of Spacecraft*, Vol. 21, No. 2.

4. SAILMAKING

Many of the design, manufacturing, and usage factors of sails are independent of ship design. As with maritime sailing ships, the sail size and shape are factors that are dependent on the design of the ship. This chapter is primarily concerned with the independent aspects of the sails.

SAIL DESIGN FACTORS

The sail is a complex structure in itself, requiring careful design from several points of view. It might be made of several layers of different materials, each of which must be precisely controlled in thickness. It might require manufacturing in a precise three-dimensional shape, with built-in billow and curving edges. In operation, the sail must be smooth on both large and small scales, without major wrinkles. It should not be made of woven materials unless the weave and fiber sizes are small relative to the wavelength of incident radiation, or the woven material is covered by a reflective layer. The solar pressure on the sail must be transferred to lines that attach to other structure without causing major wrinkles or undue stress concentrations. The sail must tolerate punctures without tearing.

Cross Section

A substrate film, shown in Figure 4.1, typically provides the mechanical strength needed to carry the loads through the sail. The loads result from the transmission of solar pressure and from packing and unfurling of the sail.

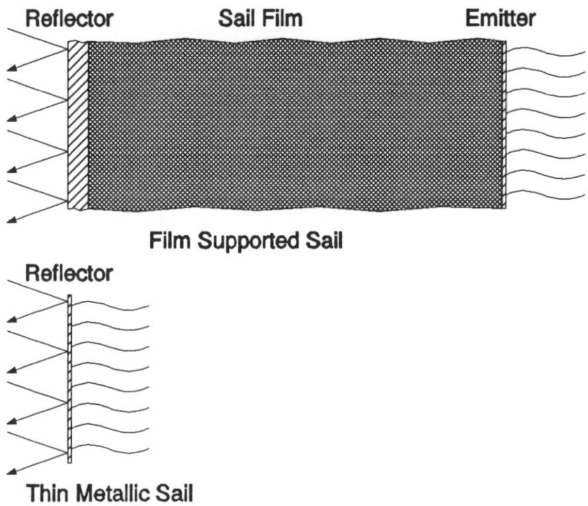

FIGURE 4.1 Sail Cross Section.

The substrate film has metallic coatings on each side. The side meant to face the Sun will typically have an aluminum coating, that may itself have an overcoating to prevent oxidation.

The substrate film might be successfully eliminated from the sail if the sail is manufactured in space and not subjected to folding. If this is done, the required thickness of the reflector might be more or less than the typical 100 nm for use with a film. The required thickness is a function of the tendon pattern used in the sail. From a performance point of view, the optimum thickness of aluminum in this application is between about 10 and 20 nm, which allows a small fraction on the incident light to pass through the reflector. An emission coating might not be needed, depending on the thermal conditions during operations. Both sides of a sail should be electrically conducting surfaces.

Plan Form

The overall plan form of a ship might be done with a single sail or with a set of smaller sails. Using a set of smaller sails is generally the best approach where there are no other practical constraints, such as a requirement for automated deployment.

Triangular sails give the minimum number of attach lines and also tend to have less induced wrinkles than other forms. An example plan form of triangular sails for the square-rigged clipper is shown in Figure 4.2.

FIGURE 4.2 Clipper Sail Plan Form.

Sail Shape

The sail will billow out under the pressure of sunlight. To minimize wrinkles, a nominal billow shape can be fabricated into the sail. The tension from billowing under solar pressure contributes to the inward curvature of the sail edges, so the edges are defined with a deliberate scallop to help minimize wrinkles. The depths of the scallop, draft, and billow and the height of the central apex above the sail corners are parameters that must be carefully defined to give the ship the desired stability and control characteristics. These features are shown in Figure 4.3.

The sail shape can change substantially with changes in the pressure on the sail. The difference between the relaxed and deflected positions for an 800 m sail under a pressure of 64 $\mu N/m^2$ is shown in Figure 4.4.

Sail Load Paths

A sail has load paths built into it to carry some or all of the tension loads. If the full tension loads are carried through the sail as a film, the areas of stress concentration could exceed the tensile strength of the film unless the film were made excessively heavy. Tear resistance is also an issue since a

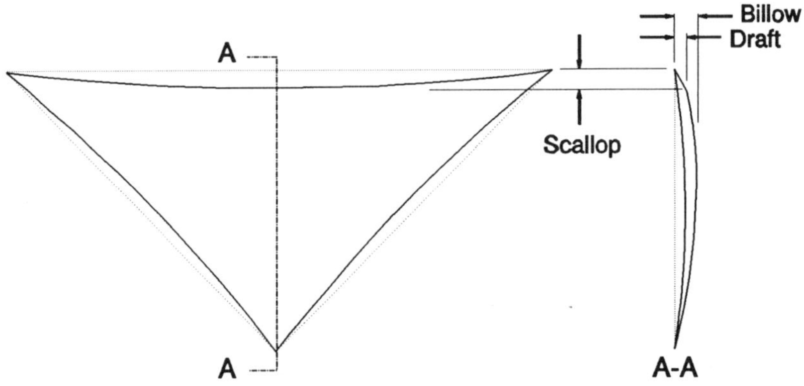

FIGURE 4.3 Curvature in the Sail.

highly stressed film will tear if it is punctured. During the lifetime of a sail, it will receive many small punctures from micrometeorites, so the stress level in the film is an important design issue.

Ripstops may be used in the sail to halt the propagation of tears. These ripstops may also serve as convenient paths for carrying tensile loads to reduce the stress in the film. A pattern of seams and ripstops is shown in Figure 4.5.

The tension carrying members in the sail are called tendons. The seams and ripstops may be the tendons, or the tendons may be separate lines that can be external to elements of the sail.

Sail attachment lines should connect to edge tendons, and to some internal tendons if practical. Spring devices in the attachment lines may be used to maintain desired tension levels in a sail. Attachment lines for a circular central opening are illustrated in Figure 4.6.

Sail Panels and Elements

An individual sail can be made in panels that are subunits of the sail. Reducing the size of the panels reduces the loads carried through the film, allowing the use of thinner film in the panels. In the limit, a plastic film substrate may not be needed if the panels are constructed in space and made sufficiently small.

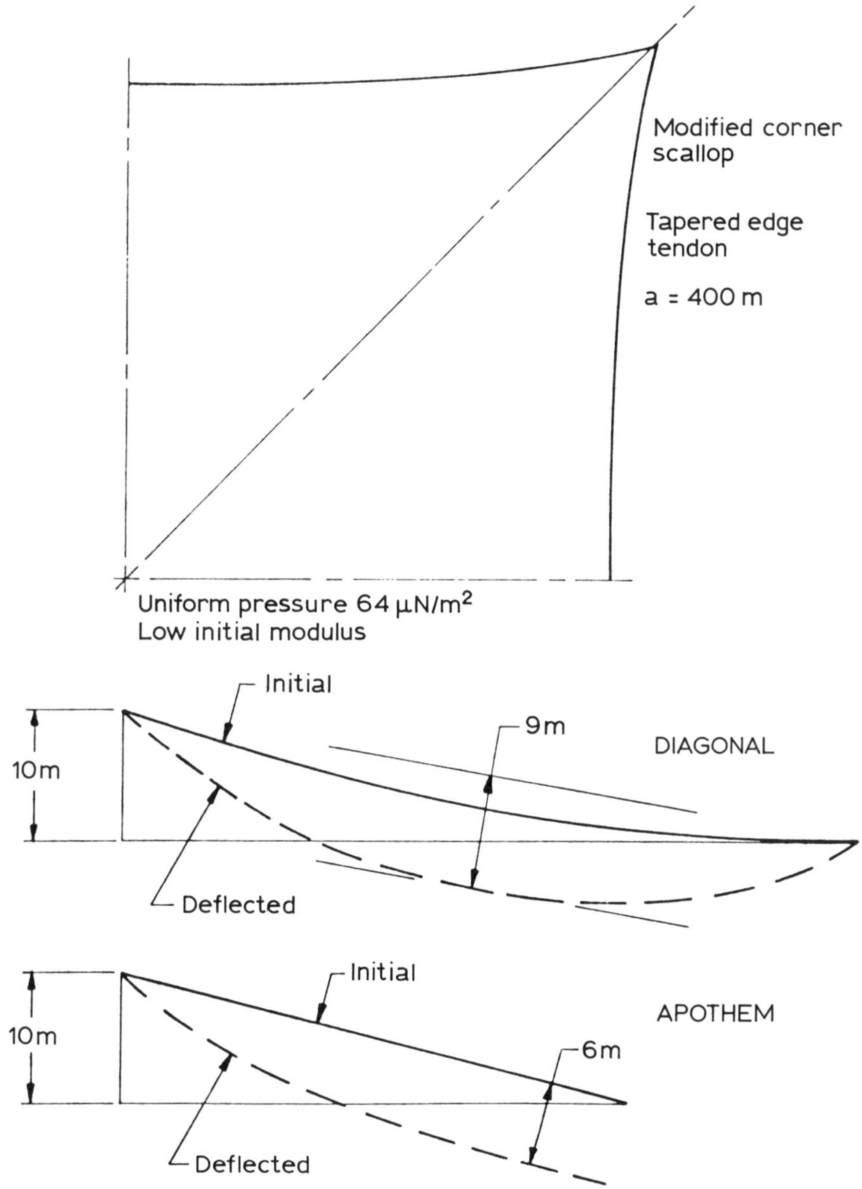

FIGURE 4.4 Load Effects on Sail Shape. (JPL)

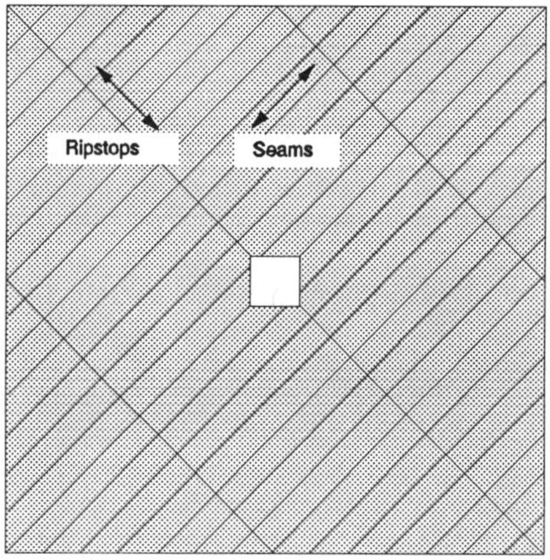

FIGURE 4.5 A Pattern of Seams and Ripstops in a Sail.

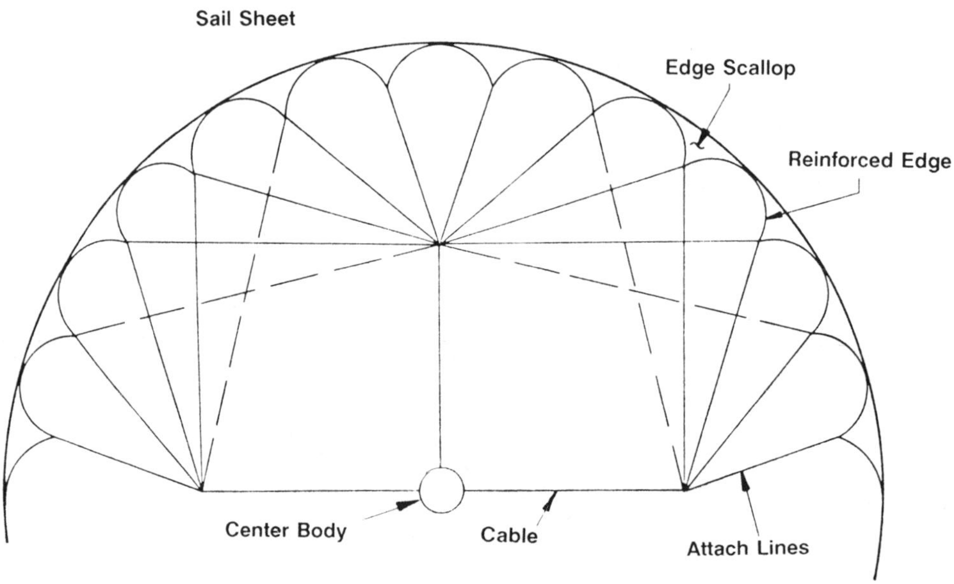

FIGURE 4.6 Sail Attachment Line Detail for Circular Opening. (JPL)

A variety of tendon patterns can be used for the panels in a sail. A sail can be made from a pattern of numerous identical panels, or from a pattern of repetitious subdivision of the sail shape.

The sail can be constructed such that it is divided into small structural elements by applying tapes across a continuous sail. The taped areas are thicker than the sail film alone, and become the paths for carrying loads through the sail. These tapes effectively divide the sail into small panels closed off by the tapes.

An alternative approach is to construct the sail as a web of tension lines. The lines are connected together where they cross, creating nodes. Small sail panels are connected to the nodes by short lines. A pattern of external tendons is shown in Figure 4.7.

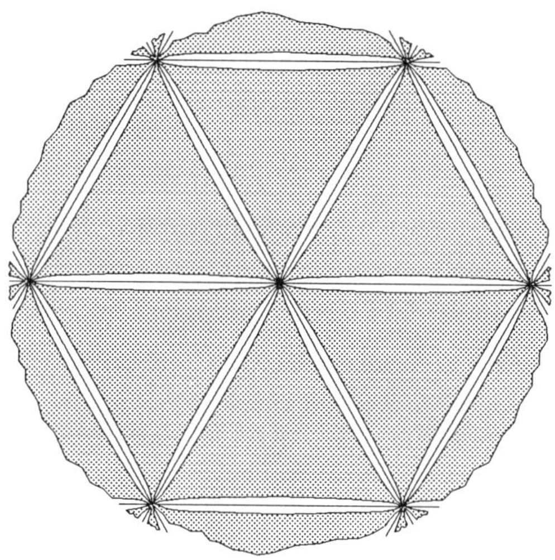

FIGURE 4.7 A Pattern of Tendons External to Sail Elements.

MATERIALS

The choices of materials and the methods of processing are important to the performance of the sailing ship. Mass, temperature limit, and cost need to be considered in making selections. Performance enhancements by some materials can have important effects on operating limits. For example, several types of coatings can by applied to the substrate to affect the reflectivity or emissivity of the sail, which can determine how close the sail can approach the Sun. The materials and sailmaking technology in this chapter is drawn primarily from the work done at JPL, assisted by several contractors, for the Halley Rendezvous mission.[1]

Sail Film

The thickness of the substrate film is important to the performance of the sailing ship since it makes a major contribution to the total mass. The maximum film thickness for even a moderate performance ship is about 2.5 μm, or 0.1 mil (0.0001 inch).

The choice of material can be driven by the expected operating temperature of the sail. Cost and availability in the proper thickness are other important decision factors.

The following sail film candidates were considered during the JPL design effort in 1976 and 1977. Technical specifications on the materials are readily available from the respective manufacturers.

Kapton [TM of Du Pont] has good radiation and high temperature properties, and was selected as the film material for the Halley Rendezvous sail. Kapton is able to operate for years at 250°C (480°F) with a tensile strength of about 56 MN/m^2 (8000 psi). At room temperature it has a tensile strength of about 147 MN/m^2 (21,000 psi). It can be used at temperatures of -120°C (-200°F). Kapton has a density of 1.42 g/cm^3. It is available at a minimum thickness of 2 to 3 μm (0.08 to 0.12 mil), and is suitable for chemical or ion bombardment etching to further reduce the thickness. Production capacity is not a problem for sail construction.

Mylar [TM of Du Pont] was used successfully in the Echo and Pegasus balloons, even though its resistance to ultraviolet radiation is very low. This weakness makes the material unsuitable for a sail in long-duration operations.

Poly para-xylylene [Union Carbide TM Parylene N] is a material that can be manufactured at a suitable thickness. However, it is not manufactured in large sheets, and its suitability for the space environment is uncertain.

The Ciba Geigy films B-100 and P-100 were found in the JPL evaluation to be less massive, less costly, and faster to produce than Kapton. However, lack of test data left significant uncertainties about their suitability for the space environment. Since the films are polyimids as is Kapton, this might not be a problem. Uncertainties also existed in scaling production up from a pilot plant to full production. The films are manufactured by Schweitzer in the US.

Some film materials are suitable for casting, which has the potential to yield very thin films. One type of water-cast film was produced in large quantities at a thickness of 0.4 μm by Midwest Research Institute. An electro-cast polyimide was made by TRW at less than 1.5 μm.

Teflon [TM of Du Pont] is not considered suitable because of its tendency to creep, or stretch, under loads.

The two leading sail film candidates in the JPL design are Kapton and B-100. Their basic properties are summarized in Table 4.1. Their mechanical properties are summarized in Table 4.2.

Reflector

The reflector is typically a vapor-deposited coating over the substrate film. If there is no plastic film, the reflective layer will also serve as the substrate for any additional layers that might be added.

Aluminum has the best overall reflectance across the spectrum, although silver and gold are similar. Aluminum usually has an overall reflectivity of 0.88 to 0.90, a specularity of 0.97, and a flatness of ±4.0. Some reflectivity values significantly above 0.90 were obtained in tests of silver and layered combinations of silver and aluminum.[2] Verification of these values and an assessment of the potential for high volume production is needed.

The transmissivity of aluminum and silver as a function of thickness is shown in Figure 4.8. The spectral reflectivity of aluminum and silver is shown in Figure 4.9. Other materials can also be used as reflectors, but aluminum is generally the best choice based on cost and full spectrum reflectivity. Another material may be better suited for a laser-driven sailer, where the illumination will be at a single wavelength.

The thickness of the reflector can be optimized for acceleration performance or it may be constrained by the need to protect the substrate from ultraviolet radiation.

The reflectivity can be further increased by using a thin film phenomenon of constructive interference by two reflective films. This is achieved by applying a transparent coating over the base reflector, followed by another reflective coating. The spacing between the reflectors, set by the thickness of the transparent layer, is one-half of the characteristic wavelength of the incident light. The spacing for sunlight is about 300 nm.

TABLE 4.1 Basic Properties of Kapton and B-100. (JPL)

Property	Kapton	B100
1. Physical		
• Density	1.42g/cc	1.16g/cc
• Thermal Expansion Coefficient (Machine Direction)	2×10^{-5}/°C between -200 to 300°C	5.25×10^{-5}/°C between -200 to 300°C
2. Thermal		
• Glass Transition, Tg	400°C	318°C
• Service Temperature Range (insitu)	-270 to 400°C	-270 to 300°C
3. Radiation		
Gamma Radiation	No problem up to 3×10^8 Rads (Si)	Not known
UV	Not known	Not known
4. Mechanical		
Tensile Strength, newtons/cm² (psi)	17,500 @25°C (25,000)	7770 @ 25°C (11,100)
	11,900 @ 200°C (17,000)	2730 @ 25°C (3900)
Modulus, newtons/cm² (psi)	301,000 @ 25°C (430,000)	252,000 @ 25°C (360,000)
	182,000 @ 200°C (260,000)	315,000 @ 260°C (450,000)
Elongation at Break, %	70 @ 25°C	10.2 @ 25°C
	90 @ 200°C	54.9 @ 260°C
Tear Initiation Strength, newtons/micron	0.2 @ 25°C	
5. Water Absorption, %	2.9	2.9

TABLE 4.2 Mechanical Properties of Kapton and B-100. (JPL)

Property and Temperature Range	Direct Produced Kapton	B100 (Made by Schweitzer)
1. Dynamic Tensile Modulus from -200 to 300°C	73% decrease	74% decrease
2. Thermal Transitions -200 to 300°C	-90°C and +120°C	-90°C and +160°C
3. Expansion Coefficient from -200 to 300°C	2.5×10^{-5}/°C M.D.[1] $3\text{-}5.6 \times 10^{-5}$/°C T.D.[2]	5.2×10^{-5}/°C M.D.[1] 5.8×10^{-5}/°C T.D.[2]
4. Ultimate Tensile Strength from -60 to 280°C	50% decrease	71% decrease
5. Static Tensile Modulus from -60°C to 280°C	20% decrease	51% decrease
6. Tear Initiation Strength from -60 to 280°C	60% decrease	40% improvement
7. Shrinkage		
8. Creep (Al/Cr* metallized) 240 to 300°C	Instrument inaccuracy prevents firm conclusion but thus far no noticeable creep has been seen	

[1]M.D. - Machine direction.

[2]T.D. - Transverse direction.

*Aluminum on one surface, ~1000A; Chromium on opposite surface, ~125A.

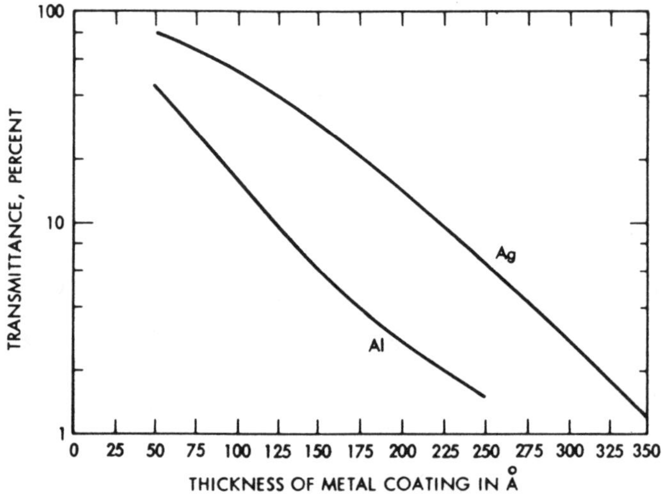

FIGURE 4.8 Aluminum and Silver Transmittance as Functions of Thickness. (JPL)

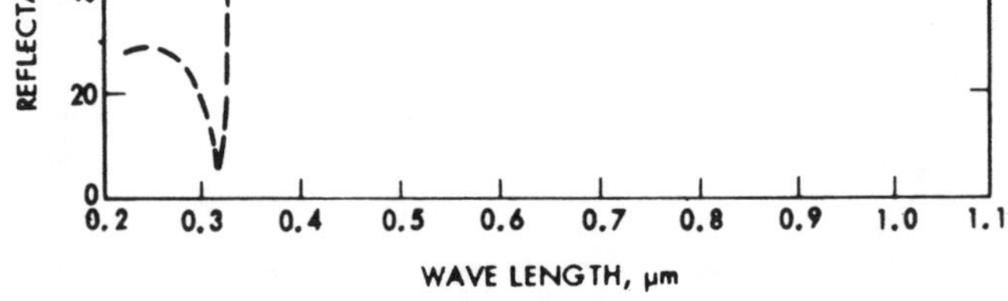

FIGURE 4.9 Spectral Reflectivity of Aluminum and Silver. (JPL)

A single thin film layer over the base reflector, for example, could boost the reflectivity from 89 to 94%. Additional like layers can be added, all of which further contribute to the reflectivity, which can be driven as close to 100% as might be desired.

The desirability of adding the thin-film layers to enhance acceleration is a matter of trading the reflectivity gain against the weight and cost increase, and considering possible materials problems such as aging and flaking. The thin-film reflective layers may be useful in allowing sails to endure very high energy densities.

Emittance Coatings

The coating on the lee, or anti-Sun, side serves two functions. It provides a means of temperature control for the sail through the value of its emissivity, and it provides an electrically conducting surface to avoid large charge buildups. Some applications might also use the backside coating as a reflector so that the ship can operate using either side of the sail. This can be done only in a limited environment to avoid thermal problems.

It is also possible to apply a coating to the front side of the sail to enhance the overall emissivity. Figure 4.10 shows a summary of some approaches to enhancing the emissivity.

Several high-emissivity coatings were identified for the Halley Rendezvous design. Chromium (Cr), tin (Sn), and indium-tin oxide (ITO) were found to be good candidates. Antimony (Sb) is also a feasible coating.

Emissivity values of 0.63 to 0.73 were achieved in tests, and higher values were thought to be possible. Temperature sensitivity of a chromium coating is shown in Figure 4.11.

The emissivity was found to have a monotonic increase as the chromium thickness was increased from 5 to 20 nm.[3] These values were obtained using film approximately 2.5 μm in thickness, and may be different for other thicknesses. The sensitivity of some emittance coatings to thickness is shown in Figure 4.12.

A possibility for achieving high emittance values is to use small whiskers on the lee side of the sail. If this is practical, the high emittance might be achieved with less mass than a coating. The techniques for production and attachment of the whiskers need to be developed.

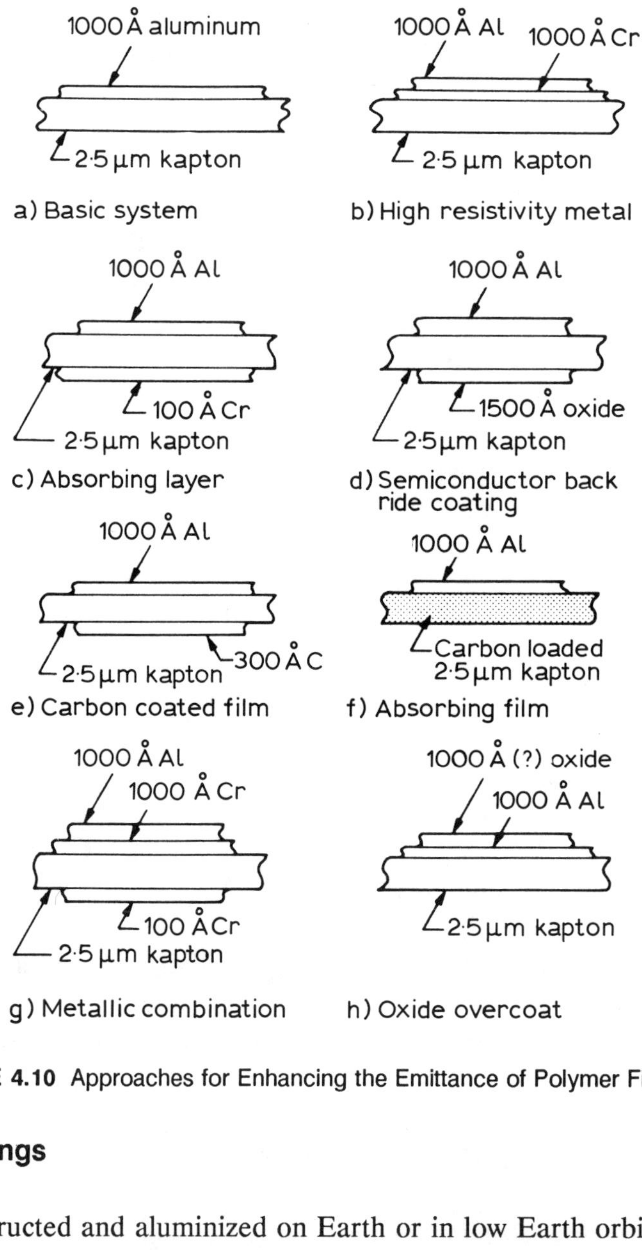

FIGURE 4.10 Approaches for Enhancing the Emittance of Polymer Films. (JPL)

Overcoatings

Sails constructed and aluminized on Earth or in low Earth orbit are subject to oxidation of the reflective metal. A thin transparent coating is applied to the reflective surface to avoid the decrease in reflectivity that would be caused by the oxidation. Some candidate overcoatings for aluminum reflectors are Al_2O_3, SiO_2, and SiO. Silver reflectors may work best with a MgF coating.

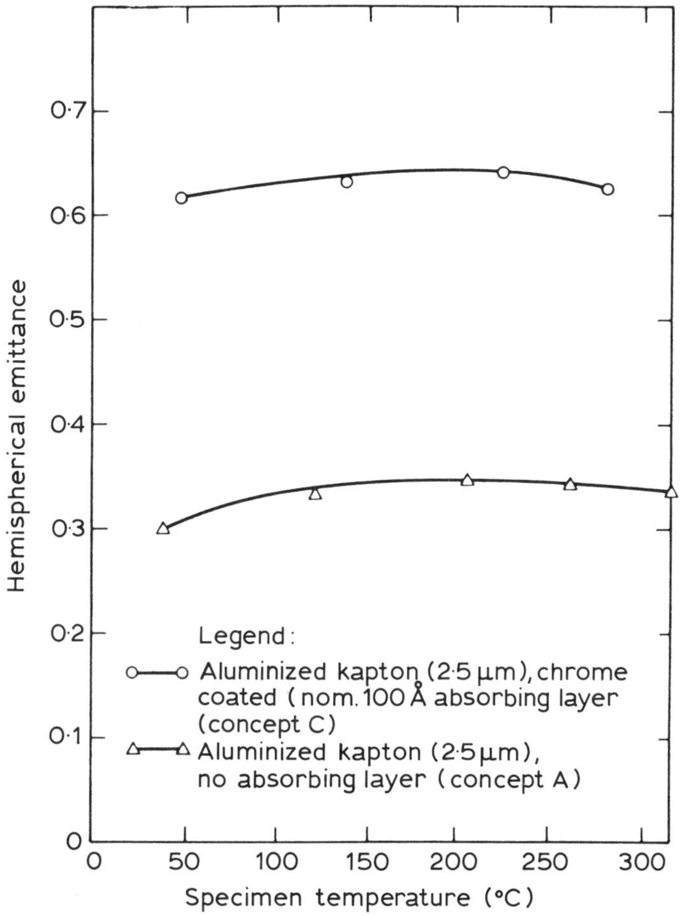

FIGURE 4.11 Emittance of Kapton With and Without a Chromium Coating. (JPL)

Seams

The Halley Rendezvous design called for seams made as butt joints with an adhesive coated tape of the same material as the sail, Figure 4.13. Tests showed that the bond lines could be kept within a tolerance of 3.8 μm (0.15 mil). The Du Pont adhesive NR 150 B2G was selected for the sail.

Overlap seams could be used if it can be shown that entrapment of light beneath film edges is not a problem or can be avoided.

Seams naturally provide a rip-stop function in the sail, and additional rip-stops can be added as desired. Both sail film and woven tapes can be used

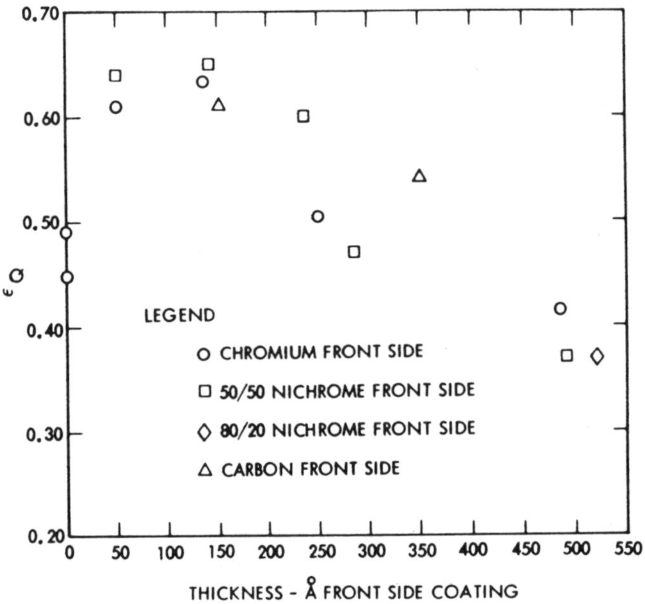

FIGURE 4.12 Sensitivity of Emittance to Thickness. (JPL)

as ripstop materials. Some types of woven tapes might also be suitable for seams.

Metallic Sail Materials

Several metals can be considered for use as metallic films, some of which are listed in Table 4.3. The table shows their mass and temperature properties.

An approach not studied to date is to use a coarsely woven, high-strength cloth, possibly of a material such as Kevlar [TM of Du Pont], that is covered by a pure aluminum film. The metallic film could be quite thin, possibly around 20 nm, since the cloth would carry the stress loads. The cloth would be cut in panels such as those shown in Figure 4.7, and attached to tendons in a sail.

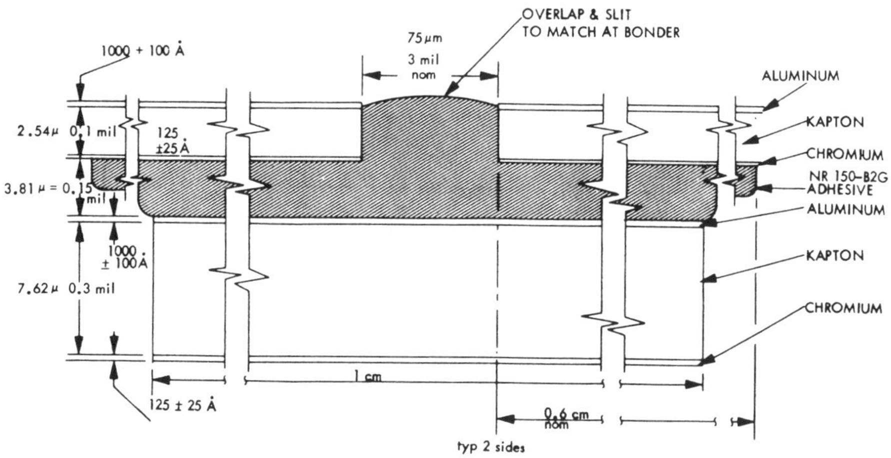

FIGURE 4.13 Butt Joint Sail Seam. (JPL)

TABLE 4.3 Candidate Metals for Sail Films.

Metal	*Density* g/cm^3	*Areal Mass at Thickness* g/m^2			*Melting Point* °C
		20 nm	*200 nm*	*1000 nm*	
Lithium	0.53	0.011	0.11	0.53	180
Magnesium	1.74	0.035	0.35	1.74	650
Beryllium	1.85	0.037	0.37	1.85	1277
Aluminum	2.7	0.054	0.54	2.7	660
Copper	8.96	0.179	1.79	8.96	1083
Silver	10.5	0.21	2.1	10.5	961
Gold	19.3	0.386	3.86	19.3	1063

SAIL FABRICATION

Film and Coatings

Commercial film production capabilities are adequate for making large solar sails. Supplementary processing, such as chemical and plasma etching, can be used to reduce the film thicknesses.

Several production techniques can be used for making thin films. The solvent casting process is shown in Figure 4.14. The water casting process is shown in Figure 4.15. A vapor deposition process, which could be used in orbit, is shown in Figure 4.16.[4]

Ion plating appears to be the best method for chromium deposition on the sail film. Existing production capabilities for coatings easily meet the needs for sail production.[5] Typical rates for application of metallic coatings are about 60 m/min (200 ft/min) of 1.8 m (72 in) film. The aluminum coating might require two passes for the full thickness.

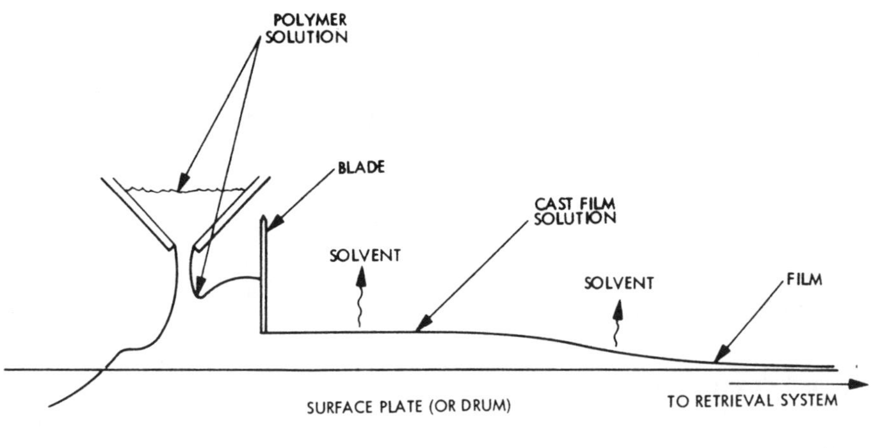

FIGURE 4.14 Solvent Casting Process for Film Production. (JPL)

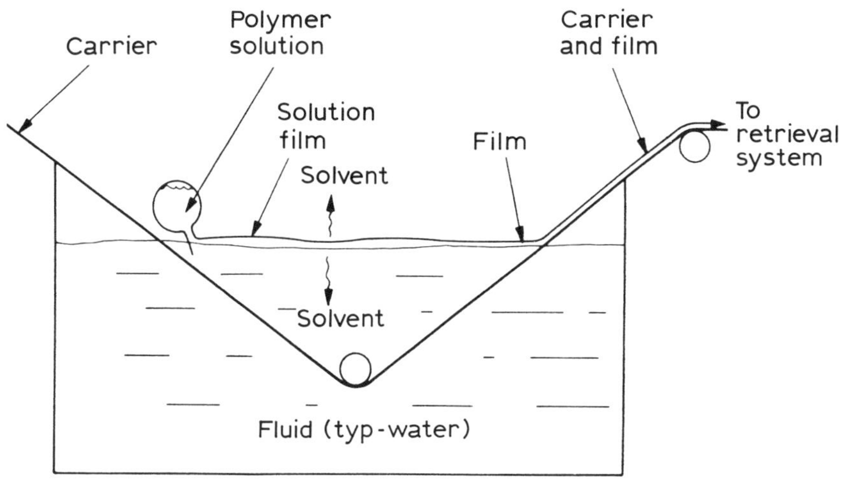

FIGURE 4.15 Water Casting Process for Film Production. (JPL)

Sail Construction

Folding during fabrication avoids the need for a full size table and reduces the potential for damage to the sail. There are two basic modes of fabrication: 1) the sail sits on the table while the seaming machine moves, and 2) the sail moves back and forth past the seaming machine. A fabricator design of the first type is shown in Figure 4.17.

Seam production rates to 2 to 3 m/min have been demonstrated.[6] The seaming task for a sail of 600,000 m^2 can be done in about 8000 working hours. This estimate is based on completing 1,000,000 m of seams, including wastage, at a rate of 2 m/min. With multiple machines and/or multiple shifts, this can be done in about one year.

The use of quarter sails on the clipper allows parallel fabrication of the individuals sails, so that all four could be in production at the same time. Likewise, several of the sails of a disk sailer could be fabricated simultaneously, reducing the overall time needed for completion.

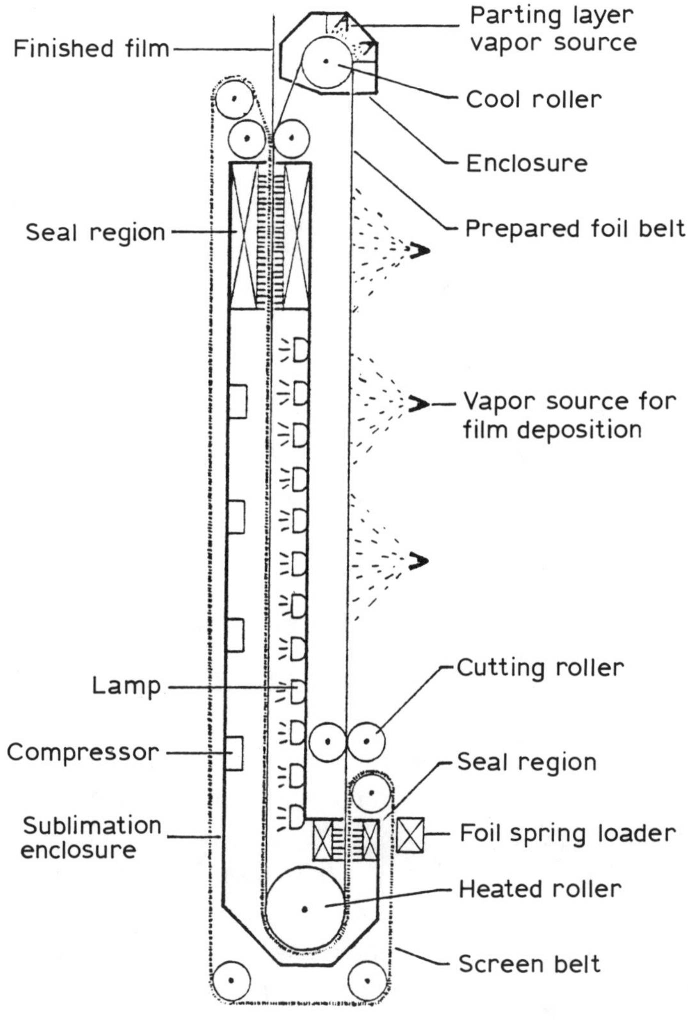

FIGURE 4.16 Vapor Deposition Film Production for Metallic Sail. (K.E. Drexler)

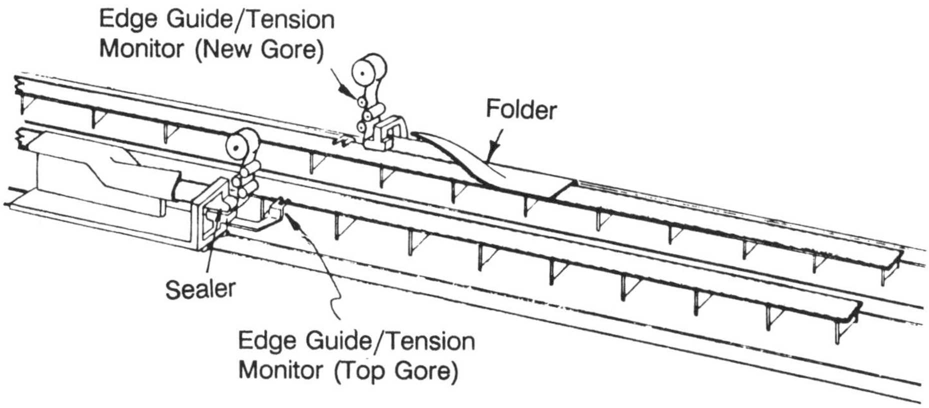

FIGURE 4.17 Sail Fabrication Machine. (JPL)

PACKING AND STOWAGE

The two basic methods of sail furling are by rolling and folding. Both packing methods are shown in Figure 4.18 for the Halley Rendezvous square sail.

The accordion fold pattern generally provides the simplest unfurling operation, while the spiral wrapped pattern avoids having double folds in the sail, resulting in the least wrinkled sail after unfurling.

The process of stowing a square sail in its canister is illustrated in Figure 4.19. The sail ribbon is folded using a special fixture so that it will fit into its canister.

Thorough venting of the furled sail is essential during and after the ascent to orbit. If this is not done, the sail could be blown out of its canister and rendered unusable.

Once a sail has been deployed, it is unlikely to ever be furled except for disposal.

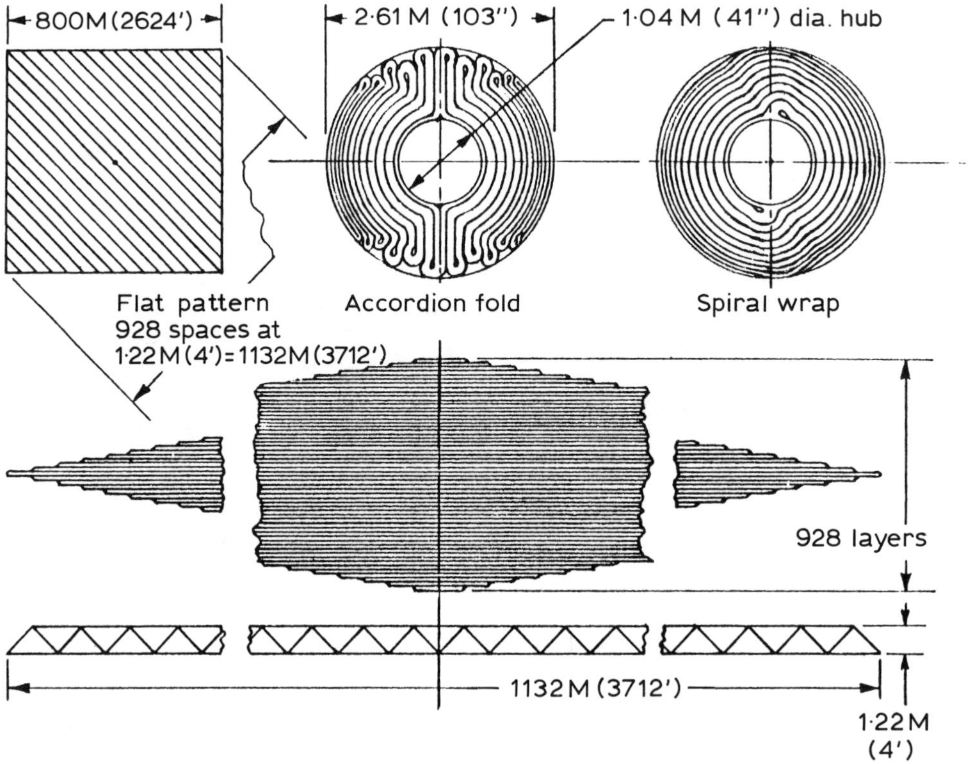

FIGURE 4.18 Packing and Stowage of Square Sail (Halley Rendezvous design). (JPL)

THERMAL DESIGN

The thermal design of a sail can be critical to its operation, especially if the sail will operate inside of Earth's orbit. The sail is so thin that its heat capacity is essentially zero, and its thermal conductance is essentially instantaneous. Sail temperature changes both with distance from the Sun and angle relative to the Sun. The size of a sail changes with temperature, and its strength changes inversely with temperature.

Thermal emissions from very thin films are not fully understood, making extensive testing necessary if the thermal design is critical. A quantum phenomenon affects the reflections and emissions from thin films, causing a deviation from classical predictions when the film thickness varies by amounts near one-half of the wavelength of the radiation.

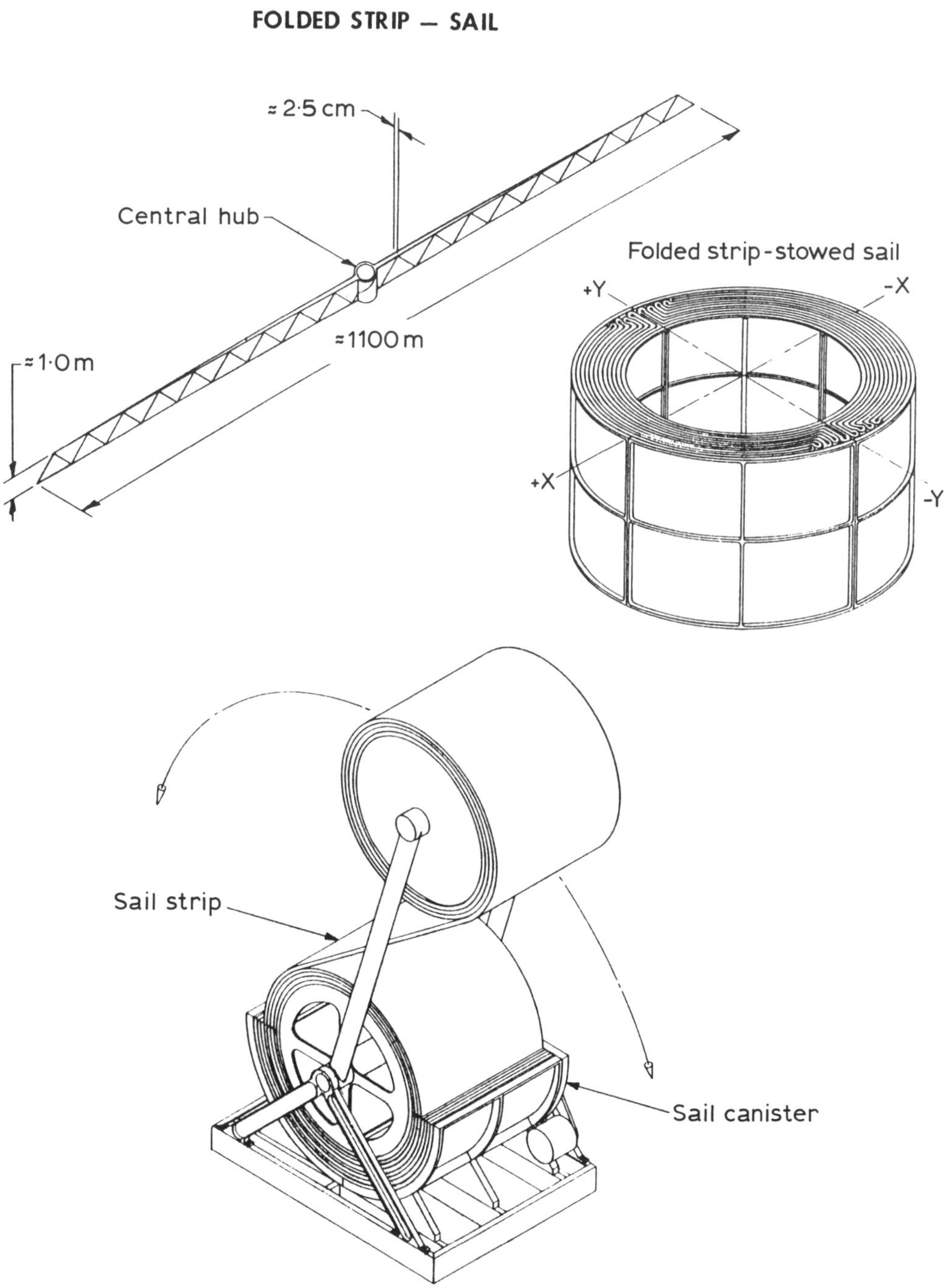

FIGURE 4.19 Stowage of a Sail Ribbon Into Its Canister. (JPL)

Since the solar radiation peak is near 600 nm and the reflector thickness is typically 100 nm or less, the reflector is much less than one-half of the wavelength of the incident radiation. The wavelength of the thermal amazons is typically in the range of 2 to 10 μm. These values are two to three orders of magnitude greater than the leeside coating of about 12 nm. However, the sail film separating the two metallic coatings is near the one-half wavelength value for the 2 to 2.5 μm film tested for the Halley Rendezvous design. In the JPL effort, testing did not extend to different film thicknesses, leaving uncertainty as to the emissivity values than could be achieved at different film thicknesses and temperatures.

For the Halley Rendezvous design, the nominal temperature at 0.25 AU, turned 30° from the Sun, is 221°C (430°F), with a maximum of 246°C (475°F), for an absorptivity of 0.12 on the Sun side and an emissivity of 0.63 on the lee side.

The variation of the sail temperature with solar distance for this sail is shown in Figure 4.20. The associated variation of yield strength of Kapton with temperature is shown in Figure 4.21.

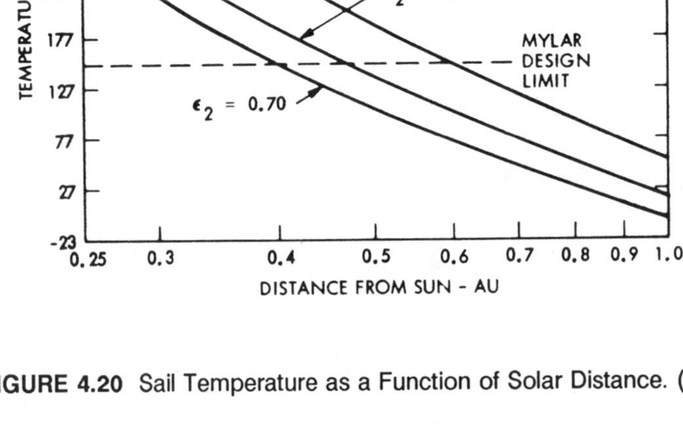

FIGURE 4.20 Sail Temperature as a Function of Solar Distance. (JPL)

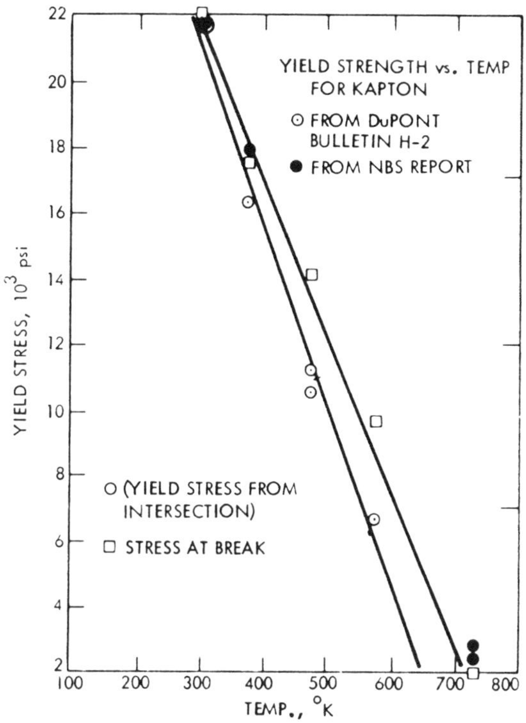

FIGURE 4.21 Kapton Yield Strength as a Function of Temperature. (JPL)

DIMENSIONAL STABILITY

Sails will have significant dimensional changes during operations, primarily due to thermal expansion and contraction. The best estimate of the expansion factor for the Halley Rendezvous design is $3.0 \cdot 10^{-5}$ m/m°C in both directions in the film, except that above 100°C, the factor increases to $5.6 \cdot 10^{-5}$ m/m°C in the transverse direction (across the film strip). The reference temperature for computations for the Halley Rendezvous sail is 0°C.

The creep behavior of Kapton is shown in Figure 4.22. The effect of creep during two years under a stress load of 70 to 140 kN/m^2 (10 to 20 psi) is less than 0.1% in both directions, although the transverse creep is slightly greater than longitudinal creep. The effect of creep is superimposed on the thermal expansion and contraction of the film.

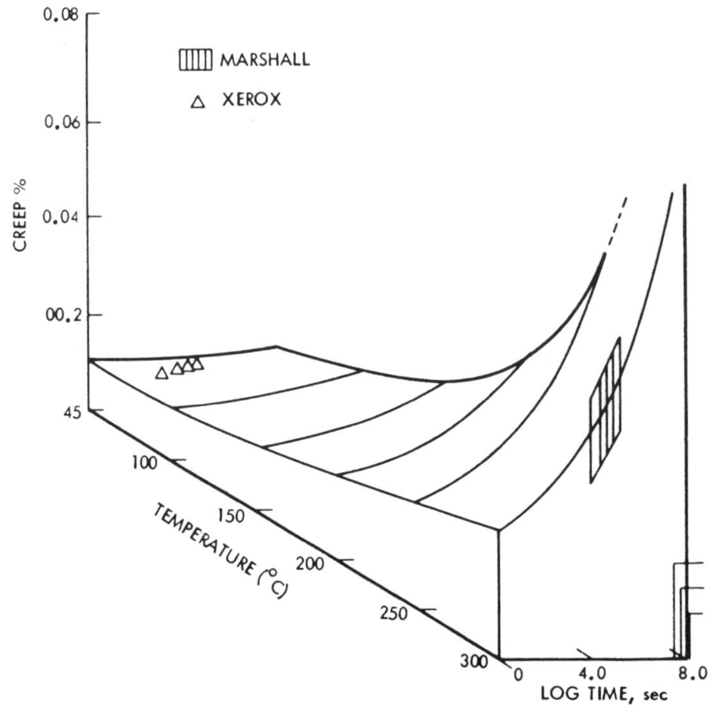

FIGURE 4.22 Creep Properties of Kapton Sail Film. (JPL)

PERFORATIONS AND ELECTRICAL CONNECTIONS

A sail can be made with perforations to reduce its mass if the holes are kept smaller than the wavelength of the incident light or microwave radiation. A pattern of such holes is shown in Figure 4.23.

The sail must be made as an electrically conductive sheet to avoid local charge buildups. As strips of sail film are bonded, some method of assuring electrical conductivity between the metallic surfaces of the strips is necessary.

Small holes can be used to electrically connect both sides of a sail where there is a continuity of electrical conduction through the holes. Each sail must be grounded to the ship, along with all masts, spars, and lines.

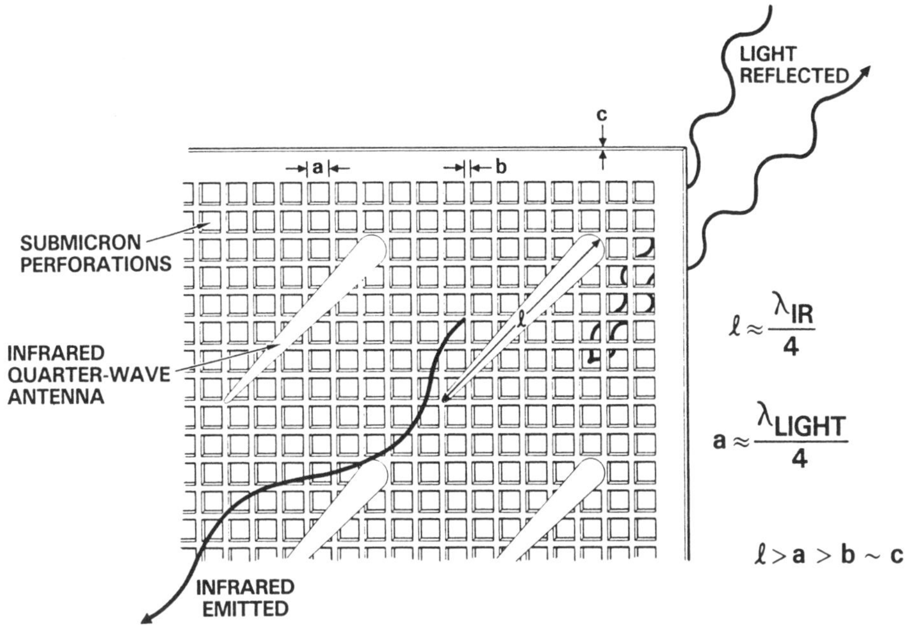

FIGURE 4.23 Perforated Solar Sail With Microstructures. (R.L. Forward/Hughes)

ENVIRONMENTAL EFFECTS

Radiation in the ultraviolet range embrittles and weakens many plastic films. This radiation can reach the film by transmittance through the reflective coating or through cracks in the coating. Plastic films should be selected with consideration given to their resistance to ultraviolet radiation.

Charged particle radiation may affect the lifetime of the plastic substrate of a sail. However, this is not thought to be a substantial factor for a sail operating in interplanetary space as close to the Sun as Mercury.

Degradation has been observed on highly reflective surfaces on spacecraft in the past. The exact mechanisms of this degradation are not known, but it is thought that ultraviolet radiation and spacecraft contamination are involved. The overcoatings earlier described have been effective in limiting the observed degradation. Careful attention to controlling the contaminating outgassing from the spacecraft should also be effective in maintaining a high reflectivity. The sail should be free of oil and other chemicals that could result in darkening when exposed in space.

Electrical charging of the sail is a result of interactions of charged particles in space. Charge buildups in the range to 10 to 100 volts have been estimated for a sail that is electrically connected front to back.

Meteoroid punctures of the sail are not expected to be very severe. Micrometeoroids will impact frequently, but the resulting damage will be confined to the local area. The size and intensity of meteoroid impacts estimated for the Halley Rendezvous mission are shown in Figure 4.24. Since most of the particles striking the sail are in the micrometer range, the lost sail area will be insignificant even over several years. Even relatively large punctures in the sail near the outer edge are not likely to exceed the control torque authority of the control system.

Sails supported by spars are subjected to low stress levels. As a result, propagation of tears from meteoroid punctures are not expected to be a

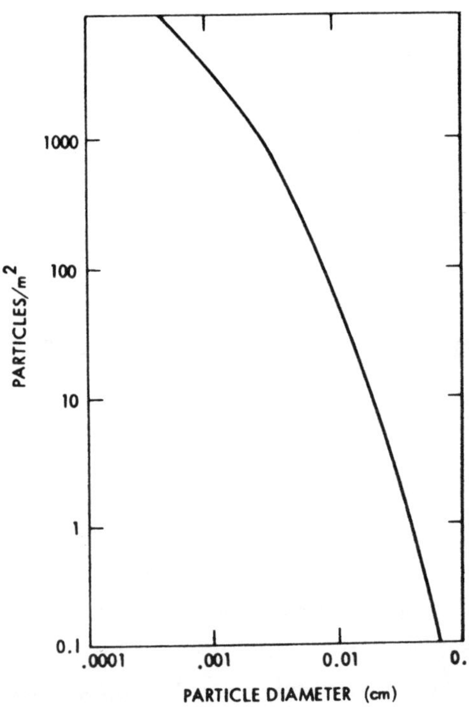

FIGURE 4.24 Number of Meteoroid Impacts During the Halley Rendezvous Mission. (JPL)

problem except possibly in extremely thin sails. In a spinning sail, the stress levels may be much higher, making it necessary to have numerous ripstop devices in the sail.

The fraction of heliogyro elements predicted to fail due to micrometeoroid damage during the Halley Rendezvous mission is shown in Figure 4.25.

MASS PROPERTIES

The substrate film can dominate the mass of the sail, Table 4.4, even when it is made thin. The mass of the seams and tendons is not a fixed fraction of

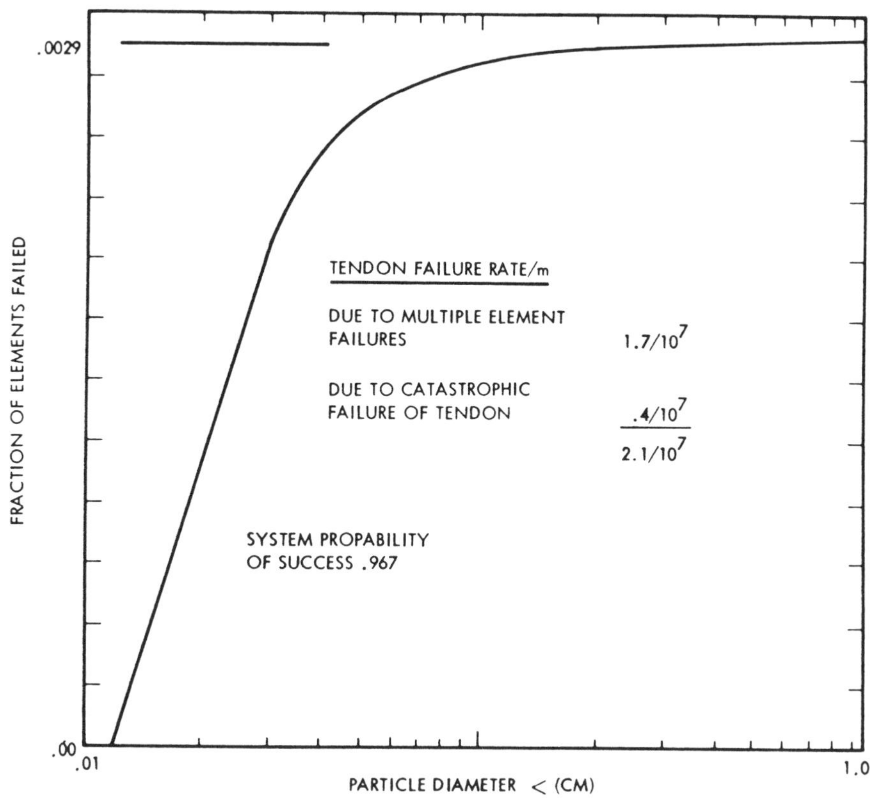

FIGURE 4.25 Fraction of Heliogyro Elements Failed Due to Meteoroid Impact. (JPL)

TABLE 4.4 Mass Properties of Various Sails.

Item	*Type*	*Thickness nm*	*Mass g/m²*
Heliogyro Sail (Halley Rendezvous design)[a]			
film	Kapton	2000	2.84
reflector	aluminum	100	0.27
emitter	chromium	12.5	0.09
joints	Kapton/Al/Cr		0.23
Total			3.43
Square Sail (Halley Rendezvous design)			
film	Kapton	2000	2.84
reflector	aluminum	100	0.27
emitter	chromium	12.5	0.09
joints	Kapton/Al/Cr		0.06
Total			3.26
Clipper Sail			
film	Kapton	1000	1.42
reflector	aluminum	100	0.27
emitter	chromium	12.5	0.09
joints	Kapton/Al/Cr		0.06
Total			1.84
Minimal Film Sail			
film	Kapton	200	0.28
reflector	aluminum	100	0.27
emitter	chromium	12.5	0.09
joints	Kapton/Al/Cr		0.06
Total			0.70
Aluminum Sail, 50% filled			
reflector	aluminum	20	0.03
tendons	Kevlar		0.06
Total			0.09

[a]Note that for the heliogyro, the sail mass is not the total blade mass. See Chapter 2.

the total sail mass, since these do not change substantially when using very thin films. As a result, thinner sails can be expected to have a relatively larger mass of seams and tendons.

The ultimate limit in areal density is expected to be that of a pure metallic sail with holes smaller than the radiation wavelength. There is considerable uncertainty in the mass of the sail tendons needed to support these panels. An ultimate lower limit on areal density for a metallic sail might be around 0.1 g/m^2 or less. In the nearer term, a practical lower limit may be around 1 g/m^2. The estimated mass properties of several types of sails are shown in the table.

REFERENCES AND NOTES

1. Jet Propulsion Laboratory (1978) *Sail Film Materials and Supporting Structures for a Solar Sail - A Preliminary Design*, Report 720-9, Vol. 4, Pasadena.
2. Endurex (1977) *Final Report for Solar Sail Project, Contract #954729*, Mesquite, Texas.
3. Endurex (1977).
4. Drexler, K.E. (1977) Design of a High Performance Solar Sail System. *MS Thesis*, Dept. of Aeronautics and Astronautics, Massachusetts Institute of Technology, Boston; (1979) *Design of a High Performance Solar Sail System*, MIT Space Systems Laboratory, Report 5-79, Boston; (1979) *High Performance Solar Sails and Related Reflecting Devices*, American Institute of Aeronautics and Astronautics, 79-1418, New York.
5. Jet Propulsion Laboratory (1978).
6. Jet Propulsion Laboratory (1978). Production trials were conducted in support of the Halley Rendezvous design. Demonstration sails have been fabricated by members of the World Space Foundation, Pasadena.

5. SAILING SHIP CONSTRUCTION

In the early years of sailing, all construction is likely to be done on Earth. A ship may be fully assembled on Earth and deployed or erected in space. Alternatively, elements of a ship may be carried into space for final assembly and deployment. Once adequate infrastructure exists in space, some or all of the construction can take place there.

GROUND-BASED CONSTRUCTION

Construction on Earth of the sails and structures for a sailing ship must meet the obvious challenges of avoiding extremely large manufacturing buildings and avoiding damage to the sails and structures during manufacturing. The work done at JPL in 1976 and 1977 included a thorough investigation of the problems of manufacturing and fabrication. The results showed that practical solutions exist at reasonable costs. The general solutions involve some method of rolling or otherwise collapsing items, and using long, but narrow, buildings to house the manufacturing processes.

Tension lines are essential elements of every sailing ship design. They can be made of either metal, such as stainless steel, or braided fibers. A tape form is generally better than a round form, because of its resistance to micrometeoroid damage and to snarling.

Tension lines can be stored on reels that pay out one or both ends of the lines. Placing them onto some type of roller during the manufacturing process eliminates any requirement for a large floor area, and keeps the items protected while being stored or moved.

Compression members are more of a challenge, but can also be handled in the same basic manner. Several types of collapsible compression structures have already been produced and used successfully in space. An example of a collapsible mast, as designed for the Halley Rendezvous mission, is shown in Figure 5.1.

Suppliers of these extendable structures include:

- Astro Aerospace, Carpinteria, CA: collapsible masts
- AEC-Able Engineering, Goleta, CA: collapsible masts
- Lockheed Missiles and Space Company, Sunnyvale, CA: wrapped ribs
- Fairchild Space Company, Germantown, MD: interlocking booms.

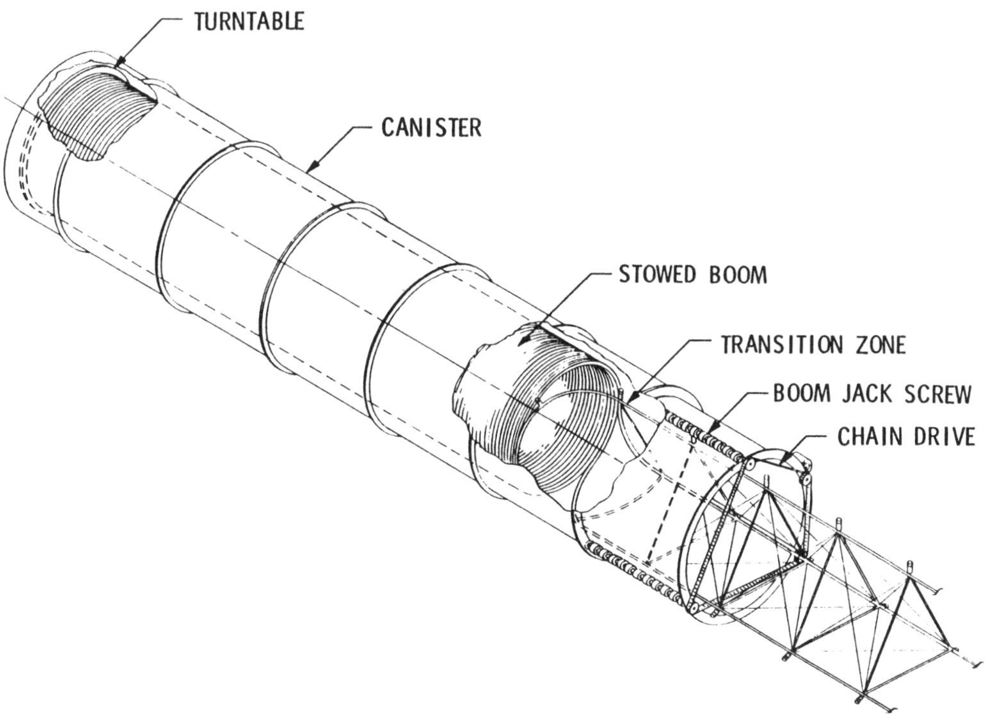

FIGURE 5.1 A Collapsible Mast For the Halley Rendezvous Sailer. (JPL)

THE SQUARE-RIGGED CLIPPER SHIP

Clipper Ship Construction

The masts and spars are supplied in canisters that can be self driving, able to extend the structure on command with their own power sources. The canisters can include rings that lock onto the booms as they extend from the canister. These rings are attach points for the stays that run between the masts and spars.

The canisters, with their masts, can be assembled on Earth and launched as a unit that does not require any further assembly in space. This is the approach that was taken with the Halley Rendezvous spacecraft.[1]

The initial packaged configuration, shown in Figure 5.2, contains the mast in a central canister surrounded by four canisters with the booms. At the base of the mast is a toroidal container with the sail. The sail is fabricated in a folded, accordion pattern creating a long ribbon that is folded or wrapped to fit into its container, as described in Chapter 4.

The numerous stays are stored on reels to help maintain control of the lines in the zero-gravity environment. The stays are made in the form of flat tapes to help prevent snarling or binding as they come off the reels.

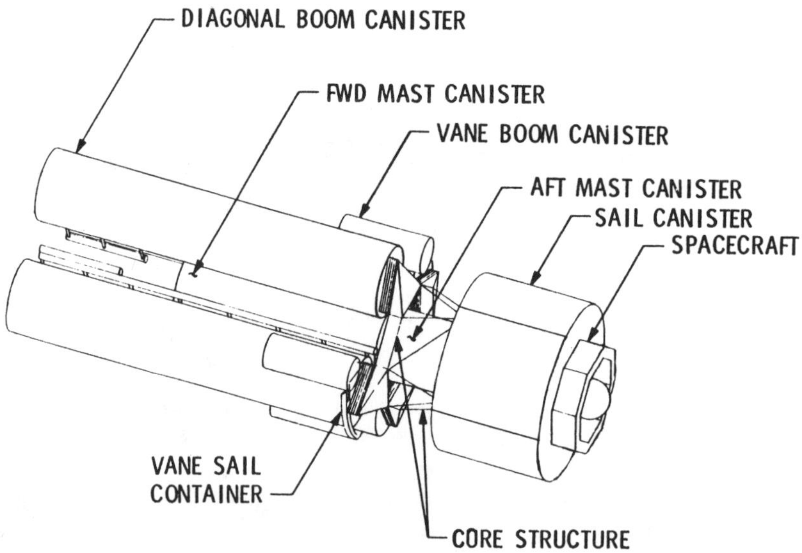

FIGURE 5.2 Launch Configuration of the Halley Rendezvous Square-Rigged Sailer. (JPL)

Clipper Ship Deployment

An automated deployment process was considered necessary for the Halley Rendezvous design because it would be done without observers and without opportunity for intervention. This requirement for automation lead to some specific design approaches in detail, but did not substantially influence the overall sail design. The steps in the automated deployment process are illustrated in the following figures. A deployment process done at a construction station in orbit will follow the same basic procedure.

The entire structure is deployed before the sail is unfurled. The steps in the mast deployment are shown in Figure 5.3. The lower mast is extended first, the boom canisters are rotated into position, then the upper mast is extended. Details of the intersection of the mast segments and the booms are shown in Figure 5.4.

The four booms are extended in a coordinated manner as the boom canisters are driven out from the central mast. The stays are pulled out by

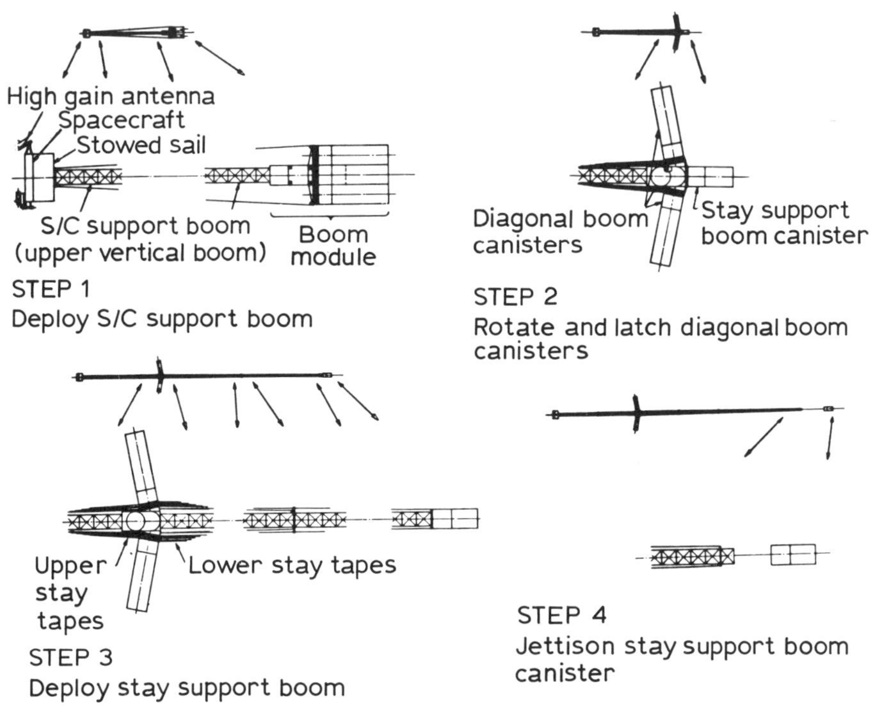

FIGURE 5.3 Deployment of Clipper Mast. (JPL)

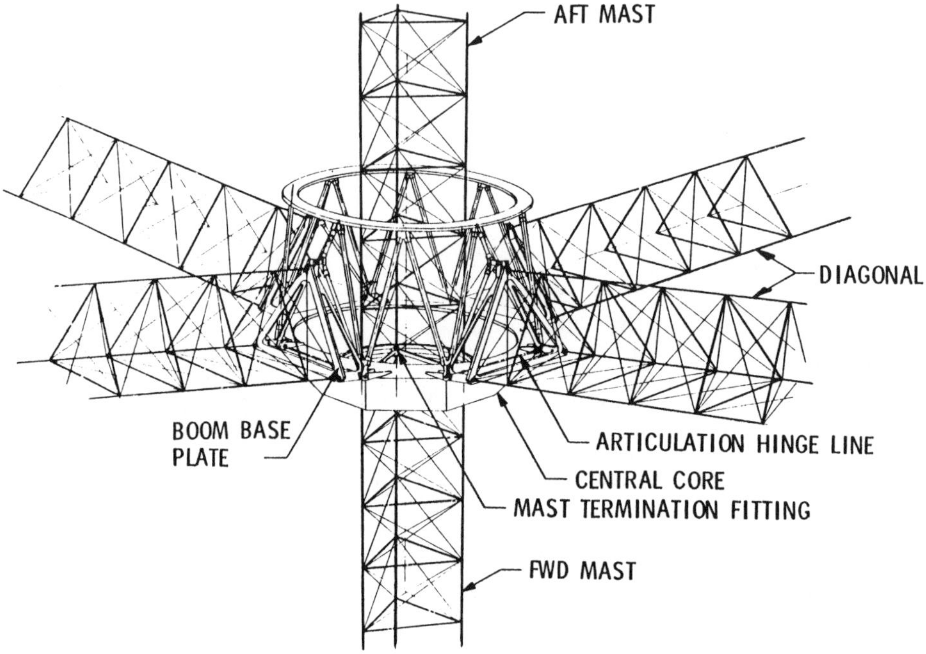

FIGURE 5.4 Core Structure Detail. (JPL)

the extension of the booms. They are attached to rings that lock onto the booms as shown in Figure 5.5. The motion of the canisters pulls the stays off of their reels by causing the reels to turn, Figure 5.6.

At each node where the stays attach to the booms, each canister pauses until all four have reached their respective nodes and transferred the appropriate rings to the booms. This assures that the structure is stabilized before the next span of boom is extended. The process is illustrated in Figure 5.7, where the canisters have moved beyond the second nodes. Each canister drives itself out to the full extension of its boom, at the final node, as illustrated in Figure 5.8.

The sail deployment sequence begins with the sail being pulled out as a ribbon to extend across the full diagonal of the structure. The next step is to pull the sail open along the remaining diagonal, Figure 5.9. This step completes the sail deployment process. If the sail is made in quarter panels, each panel would be pulled open in a similar two-step process. During the sail unfurling operation, the ship should be oriented so that sunlight does not directly enter into the folds of the sail to avoid possible thermal damage.

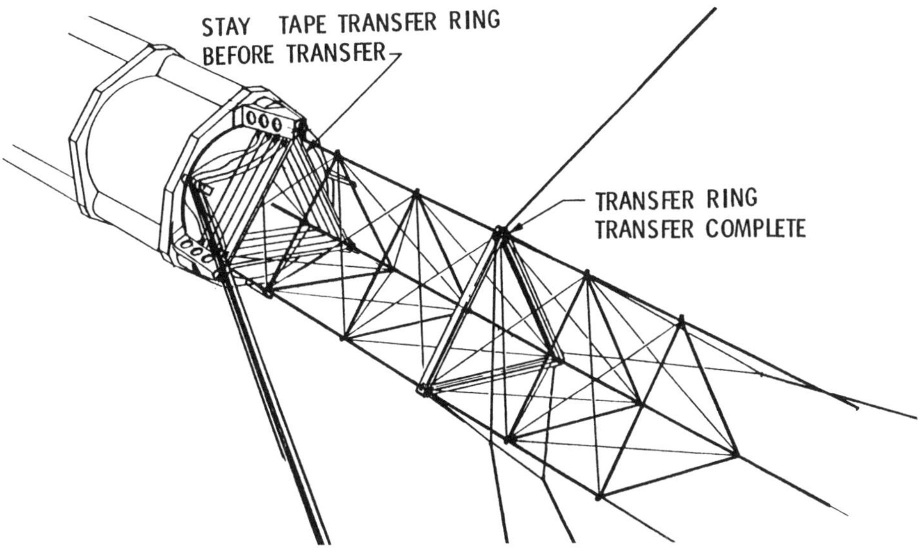

FIGURE 5.5 Boom Stay Ring Detail. (JPL)

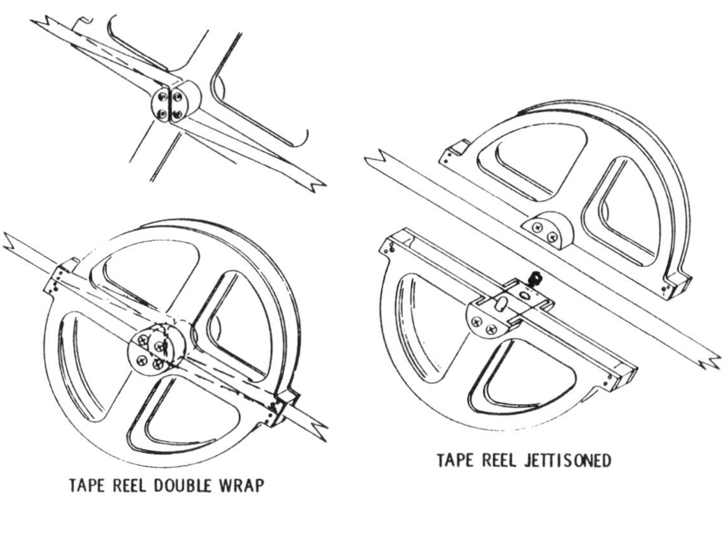

FIGURE 5.6 Stay Reel Detail. (JPL)

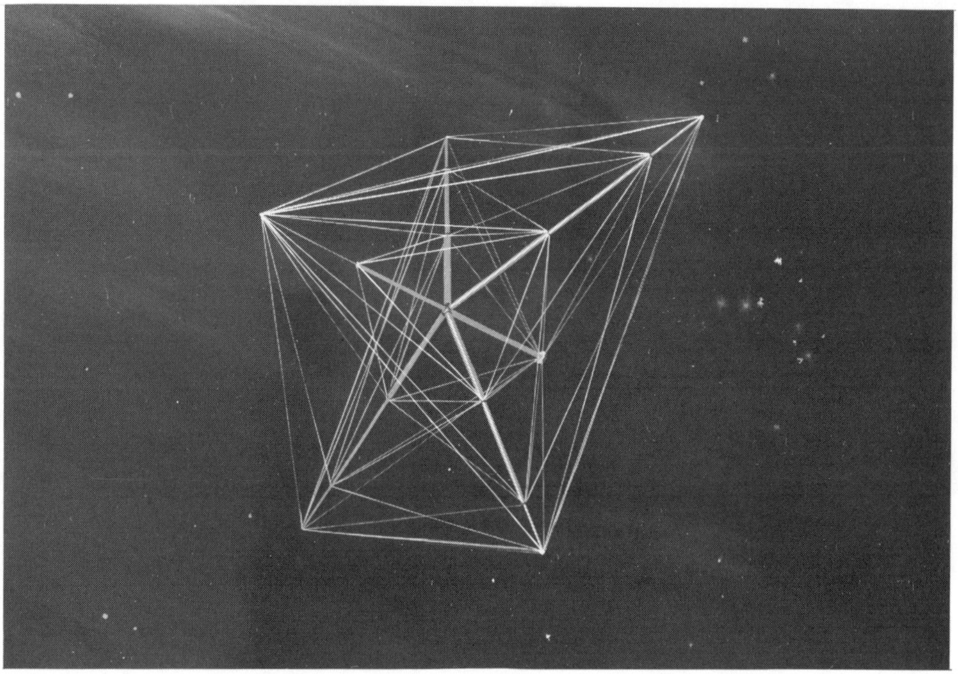

FIGURE 5.7 Deployment of Clipper Booms. (JPL)

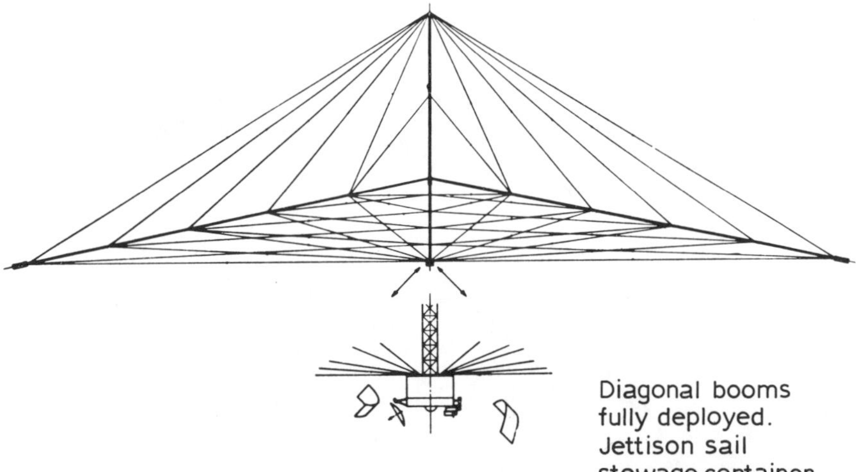

FIGURE 5.8 Boom Extension Completed. (JPL)

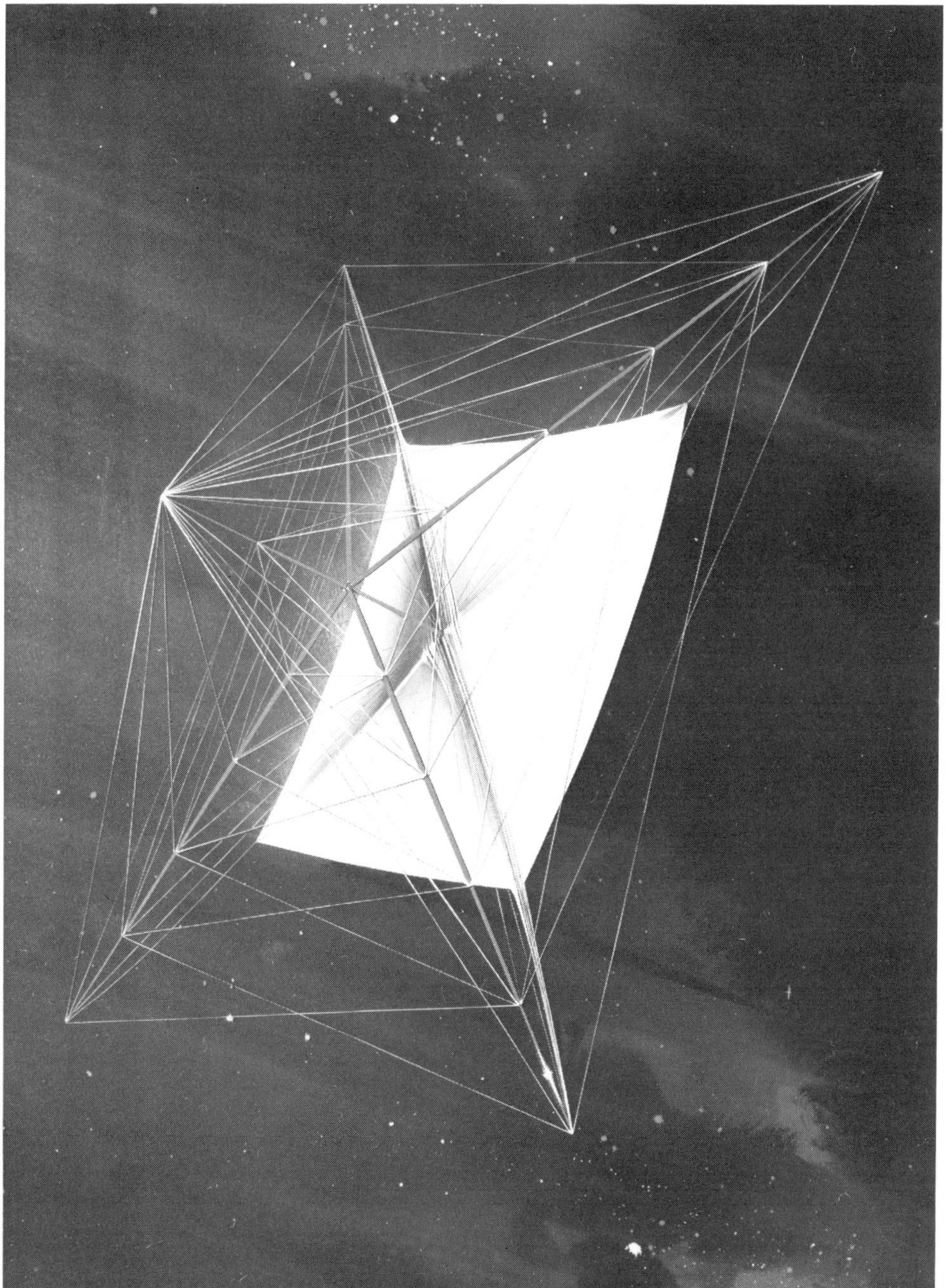

FIGURE 5.9 Unfurling of Clipper Sail. (JPL)

If the ship uses vanes, they are deployed as the next step. The vanes are themselves smaller solar sails with their own supporting structures. The deployment of the vanes is a process like that of the main sail. A deployed vane is shown in Figure 5.10.

The boom canisters, now located at the outer boom tips, are driven off the ends of the booms to remove their dead mass. All of the lanyards for holding and controlling the sail are set into place as the boom canisters move out along the booms in the extension process.

If this type of ship were erected in low Earth orbit, workmen could observe the process and make corrections. They could even be an integral part of the erection process, removing the requirement to develop a fully automated process. With manual assistance, the basic deployment or erection process would be much the same even though the details of implementation might be substantially different.

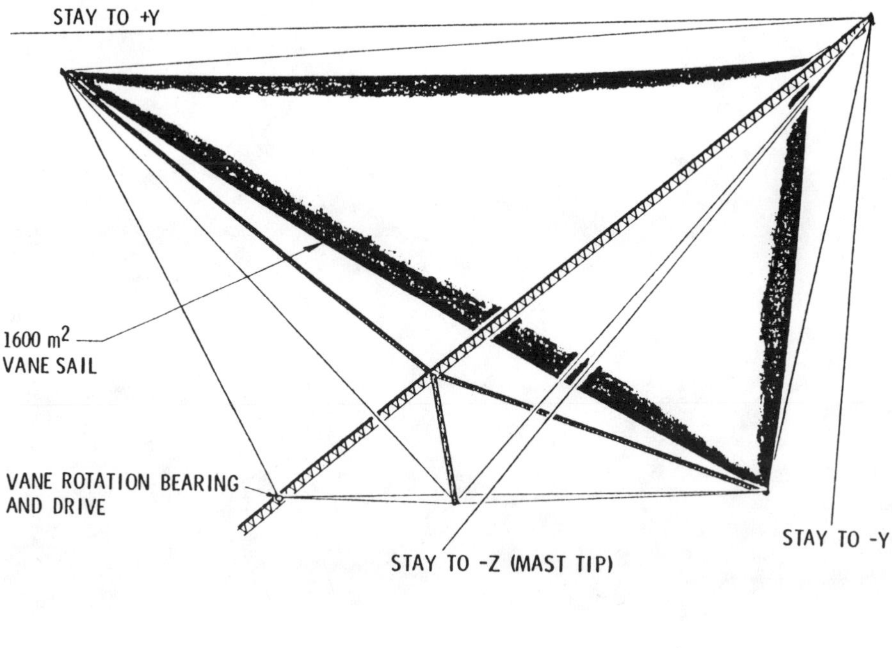

FIGURE 5.10 Deployed Vane Detail. (JPL)

THE WRAPPED-RIB SAILER

Wrapped-Rib Sailer Construction

The nature of this design requires that the ribs be simultaneously wrapped around the core, and also simultaneously unwrapped in space. A design challenge is to avoid having the ribs and sail panels, or solid sail, fully open before packing can be done. This might not be a major problem with small sail sizes, but it would be impractical for larger sail sizes.

There appear to be conceptual solutions using individual sail panels, each folded and wrapped on a temporary drum. As the central body is rotated, the sail panels simultaneously come off their drums and wrap around the hub between their respective supporting ribs. The ribs likewise can be on temporary drums. This pattern is illustrated in Figure 5.11. A possible alternative is to have the sail wrapped around the core above the ribs.

Wrapped-Rib Sailer Deployment

The wrapped rib configuration has a deceptively simple deployment process. Stored energy in the wrapped ribs provides the deployment forces, so that

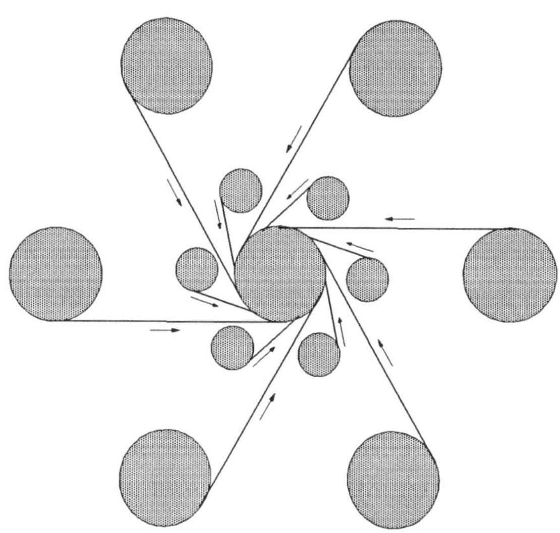

FIGURE 5.11 Stowage of Wrapped Ribs and Sail.

when fasteners are releases, the ribs unwrap themselves from the hub, pulling the sails open in the process.

The ribs must be controlled during deployment so that the rate of opening remains acceptable. The restraining mechanisms must not damage the sail or ribs during the process. The changing geometry of ribs, sails, and attachment lines provides a design challenge to avoid interference and snarling of the components.

The process of unwrapping the ribs opens the sails since the sail perimeter is attached to the tips of the ribs. The individual sail panels will have attachment lines to the ribs. Some or all of these lines may be adjustable to provide a means of changing the positions of the individual sails.

THE DISK SAILER

Disk Sailer Construction

The disk sailing ship uses flexible tension structures except for the central hub and rim, allowing the sail support structures to be stowed with the sail itself. This requires that the structural elements within the sail to be fastened together with the sail material as the sail is stowed.

The method of deployment and the method of giving the sail its initial spin constrain the methods of manufacturing and stowing the sail. It appears advantageous to manufacture the sail in gores that fasten together along their edges after they have been stowed. This can be done using coded tabs extending from the gore edges. After the stowed gores are put in place, the lines linking the corresponding tabs can be attached.

The gores can be made on tables and folded or taken up on rollers as material is added. The stowed gores and their restraining structure are attached to the rim, which may be done when the rim is fully folded. This allows separate fabrication of the hub and rim unit.

The rim itself is made in hinged segments, with each segment attached to the hub by pairs of lines. The hinges allow the segments to fold 180°, so that the rim segments lie tight against the hub in their stowed positions. A sail gore will be attached to each rim segment. The fully stowed configuration is shown in Figure 5.12.

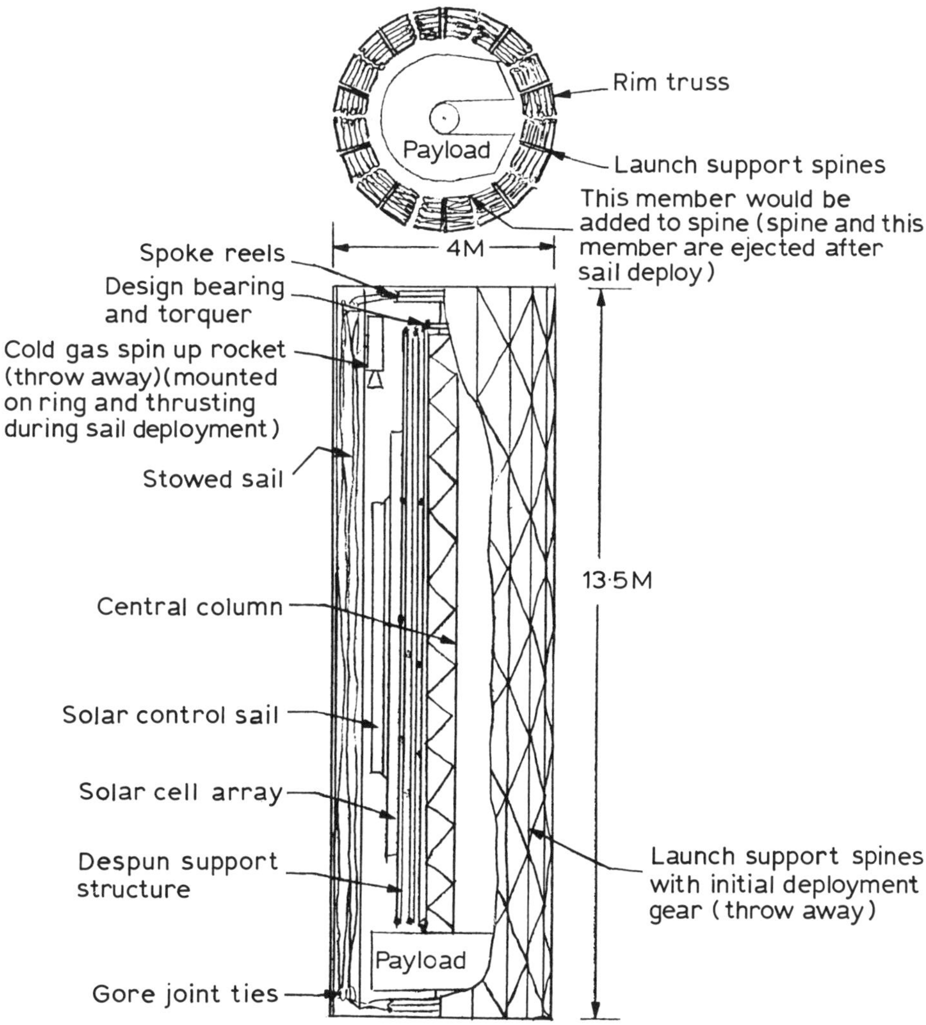

FIGURE 5.12 Disk Sailer Stowed Configuration. (JPL)

Disk Sailer Deployment

Since the disk sail has no structure that can give it a definite shape, the ship must have some amount of initial spin so that the centrifugal acceleration will hold the sail in place. The initial spin rate is provided by a source such as small chemical thrusters. This spin must be sufficient to begin extension of the sail elements, but not so great that the sail would be damaged. The sail elements themselves must then begin providing the angular momentum increase to enable the remainder of the sail unfurling. The unfurling pattern must be compatible with the use of solar pressure on the sail panels to increase the ship's angular momentum during unfurling.

Deployment of a packaged, prefabricated disk sail and its structure is a complex process, one which was not fully solved during the design effort at JPL. The process identified in that effort is described here.[2] The first step in the deployment is the extension of the hub rim, done under a moderate rotation rate for the vehicle. This deployment is done by paying out the restraining tape spokes while the rim unfolds in a *Z* pattern. An intermediate and the final positions are shown in Figure 5.13.

Once the hub rim is fully opened and the spokes are fully out, the sail gores can be deployed. The centrifugal effect provides the deployment force needed to pull the sail gores out. They must deploy in a controlled, synchronized manner without being prematurely pulled open by the centrifugal effect. Some methods of gore deployment are shown in Figure 5.14. The folding/unfolding pattern was not thoroughly explored, and may contain some substantial problems.

Unfurling sail panels from full-width storage rollers may be an attractive approach to deployment. The sail panels will not have creases or serious wrinkles, and will not be subjected to large deployment forces because the rollers are constrained. This process could be especially attractive for ultra-thin sails.

Structural tension lines can be build into the panel edges, or can be deployed along side the panels off the same rollers.

The design rotation rate is 0.3 revolutions per minute for the JPL design. Solar pressure vanes or rotation of the gores are likely to be the best means of adding angular momentum during the deployment process.

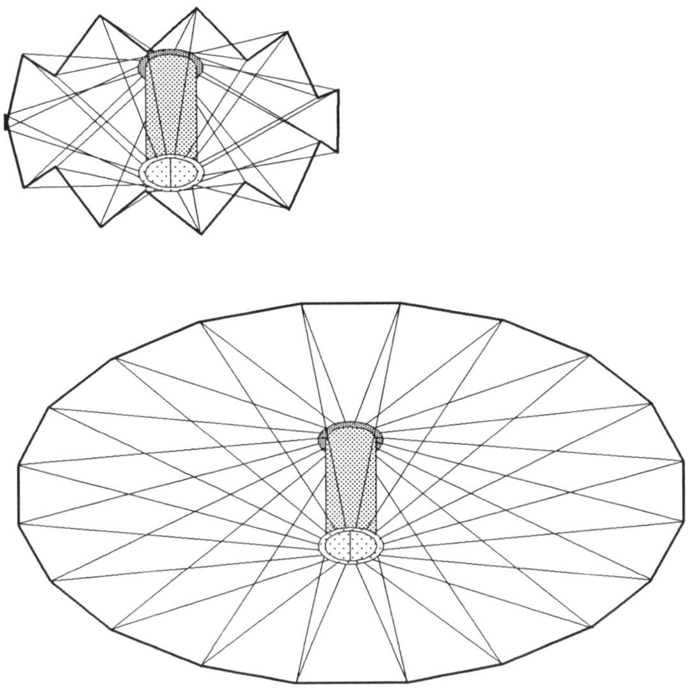

FIGURE 5.13 Deployment of the Disk Sailer Central Structure.

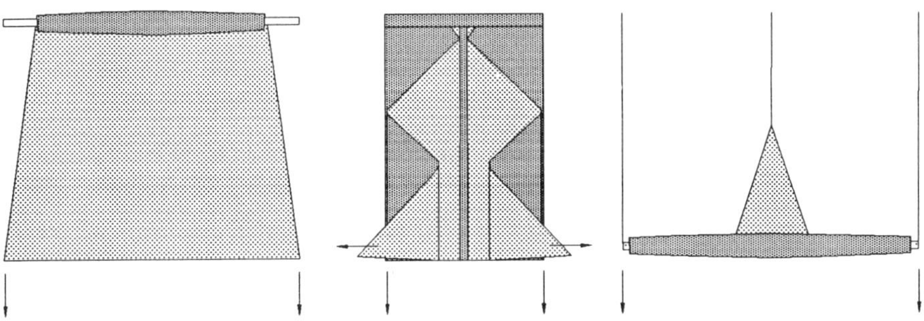

FIGURE 5.14 Methods of Unfurling Disk Sail Gores.

THE HELIOGYRO

Heliogyro Construction

Construction of the blades as integral units is the largest part of the construction effort. The sailmaking process provides the blades complete and stowed on their rollers, including the edge tendons.

The balance of the construction process includes the hub and the blade attachment mechanisms, done with conventional construction techniques. The completed unit includes the blade rollers mounted and folded against the hub.

Heliogyro Deployment

The simplicity of the deployment process of the heliogyro is its most attractive feature.[3] The first phase is the deployment of the rigid structures as shown in Figure 5.15. The arms for the blades extend outward to lock into place, then the roller assemblies are rotated into position. At this point, the sail blades are ready for extension.

The heliogyro uses the centrifugal effect for blade deployment as shown

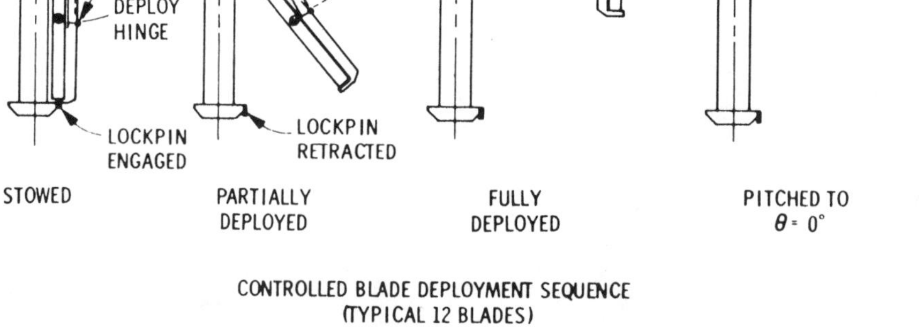

FIGURE 5.15 Heliogyro Structural Deployment. (JPL)

in Figure 5.16. An initial spinup of the vehicle is done by chemical thrusters, then the rollers are allowed to pay out the blades in a synchronized pattern.

During this process, the axis of the hub is aligned with the Sun and all rollers are held at the same collective pitch angle. This enables use of solar pressure to provide the necessary angular momentum. The unrolling process continues until the blades are fully extended and the blade stabilizer bars are extended. The rate at which the blades are unrolled is limited by the rate at which angular momentum can be added, Figure 5.17. The Halley Rendezvous configuration requires nearly two weeks to complete the deployment process.

SPACE-BASED CONSTRUCTION

The construction of sailing ships in space has not been analyzed as yet in any substantial depth. The technology has obviously not been developed, but

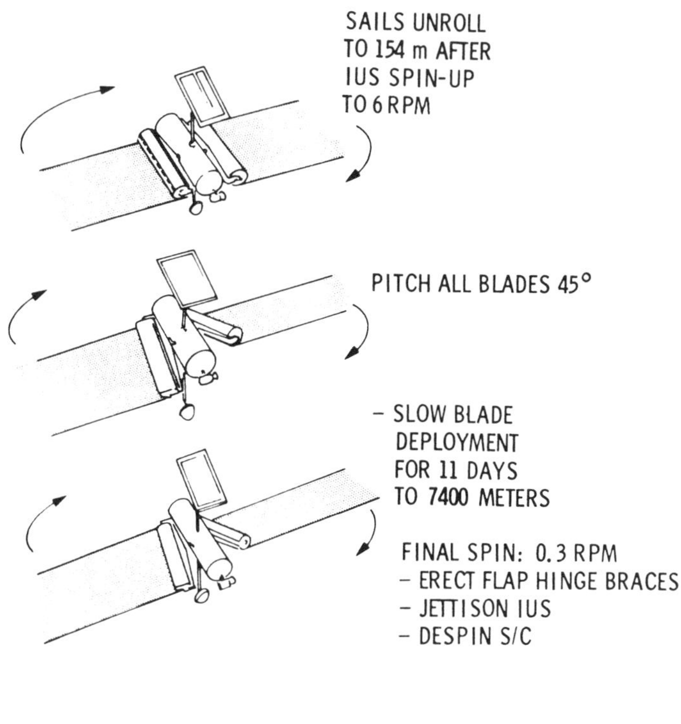

FIGURE 5.16 Heliogyro Unfurling Process. (JPL)

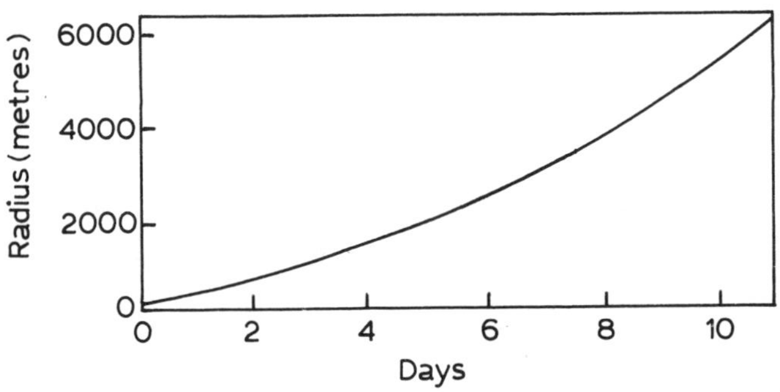

FIGURE 5.17 Heliogyro Blade Deployment Schedule. (JPL)

there do not appear to be any necessary tasks that cannot be done in space.

Reduction, even elimination, of direct manual involvement is desirable, so there are important benefits from using automation and telerobotics in the construction process. Workers and crews in pressure suits will not be able to pull the fragile lines and sails in the manner of maritime sailors. Space sailors will need to be more indirect.

Ship construction in space will be done from some type of construction base. This might be a general-purpose station used for a variety of other jobs, or it might be a smaller base dedicated to the construction of large space structures, or even specialized for just sailing ship construction. The base may be located at low altitudes if tethers are used to deploy the ships when they are completed. This procedure is discussed in the next chapter.

Construction would be easier at bases located at higher altitudes where gravity gradient effects are less and aerodynamic drag is negligible. The operations costs are likely to be greater with a higher altitude base because of the propellants required to transfer between low orbit and the base.

The first ship construction jobs seem likely to be more economical at a low-orbit station, assuming they are done when there is still a modest infrastructure in space. In the presence of extensive sailing operations, a dock might be established farther out from Earth. Once that is done there would be a choice of where to do ship construction, but economics may still favor a low altitude site for construction.

Film and Coatings

The substrate films can be processed and coated in space, probably in much the same manner as on the ground. The film can be stored in rolls, to be taken off during the process of fabricating sail elements. Thin films can be handled in this manner.

Sails made of pure, ultra-thin aluminum may require handling only in support frames. The panels would be transferred directly from the frames into the lattice network of the sail.

Rigid Structures

The square sail design developed for the Halley Rendezvous mission relied on an automatic deployment process that resulted in a rigid structure when completed. The study showed that a rigid structure can be erected in space without need for a temporary structure to stabilize the ship structure during erection. However, this required careful attention to each step of the erection process to assure that the partial structure would not collapse.

A partially erected structure does not have all of the load paths in place to provide full stability of the structure against collapse. An external scaffolding can serve as a template against which the ship can be constructed. This scaffolding can provide the load paths to support and stabilize the new structure during the erection process. The use of a scaffolding can reduce the risk to the new structure and simplify the construction process. A substantial reduction in the ship's structural mass is possible due to elimination of the requirement for the structure to carry the deployment loads.

Other factors must also be taken into account, however. The scaffolding may cost more than the construction savings resulting from its use, especially considering the cost of transporting the scaffolding into space. The scaffolding may also be expensive to maintain between construction jobs. The attached scaffolding may preclude using the base for other purposes unless the scaffolding is taken apart or allowed to drift free, which may not be desirable.

The decision to use scaffolding or to design the ship structure to be self-stabilizing during construction will be a part of any sail construction project. Cost and risk, as always, will be important deciding factors.

Sail Construction

Sail fabrication in space could be done with rolls of film from Earth using a sailmaking frame. Sail elements constructed in space can be furled onto rollers or attached to rigid frames to avoid creases or damage while being transported.

When extremely thin sails are used, it might be necessary to manufacture the films in space to minimize damage from handling. Thin plastic films manufactured in space could be placed onto rollers, and transferred to a sailmaking frame to build up sail elements. Metallic sails, lacking a structural backing, may need to be made directly in a frame.

Sailmaking frames could be used on a wide variety of scales. They could be used for sail elements only a few meters across, or they could be large enough to hold the entire set of sails for a square-rigged clipper, a disk sailer, or a lattice sailer.

The Square-Rigged Clipper in Space

A square-rigged clipper built in space could potentially have a mass substantially less than one built and assembled on Earth. The masts, sails, and lines of a square-rigged ship, or other spar-supported ship, may be manufactured on Earth and then assembled in space. The masts and booms could also be manufactured in space in configurations that are not suitable for stowage in canisters, with potentially better mass and stiffness properties.

A possible construction platform configuration is shown in Figure 5.18 that may be suitable for clipper ship construction in low orbit. The platform, dwarfed by the ship, serves primarily as a habitat, machine shop, storage depot, and massive anchor for the ship. The platform could include a full-sized construction frame that serves as a construction template and also carries all structural loads until the completed ship is released.

The Disk Sailer in Space

The construction of a disk sailer in space could be done with scaffolding attached to the central structure of the ship. The ship and platform would likely remain together until ship construction is completed.

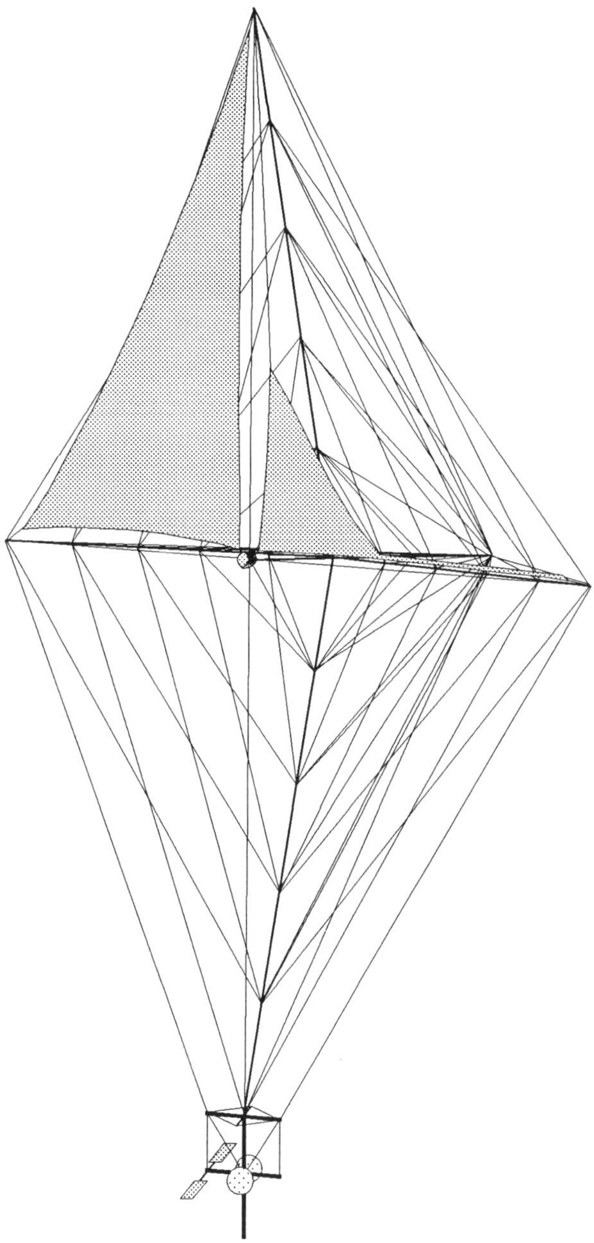

FIGURE 5.18 Sailing Ship Under Construction at a Platform.

In one approach, the construction frame would be nominally the size of the central structure, about 10% of the diameter of the ship. The ship's rim could be made as a solid, low-mass structural ring without the segmenting and hinging needed for a structure built on Earth. The construction frame carries all deployment loads until the ship is completed. The sail panels are deployed when the ship is rotating.

Another method of construction is to use a full-sized frame to hold the radial sail members while the sail is being completed in space. This approach allows the structural tendons and sail panels to be put into place as separate items. The frame will need to have a modest rotation rate before release of the sail so that the sail will have some degree of stiffness.

A third approach is to construct the sail without a frame while it is rotating. Techniques can probably be developed that could make this approach practical. This approach might have the lowest cost, especially for larger sails.

A key consideration is to avoid the permanent presence on the ship of masses, such as sail rollers, that are only associated with construction and deployment. Carrying this mass for the duration of the ship's operating life is very undesirable.

THE LATTICE SHIP

The lattice ship lacks the rigid central structure of the disk sailer, and also has the intricate lattice network of lines that must be kept under continuous tension. As a result, the ship appears to be amenable only to fabrication in space, and possibly only in a construction frame.

A scaffold structure has been proposed by Drexler.[4] A cross section of the scaffolding with a ship inside is shown in Figure 5.19. An external view of the structure is shown in Figure 5.20.

The ship is constructed with the frame at rest if it is a rigid structure, but the entire frame and ship must be given a modest rotation rate before the ship is released. If the frame is not rigid, it must be in continual rotation to maintain its stiffness.

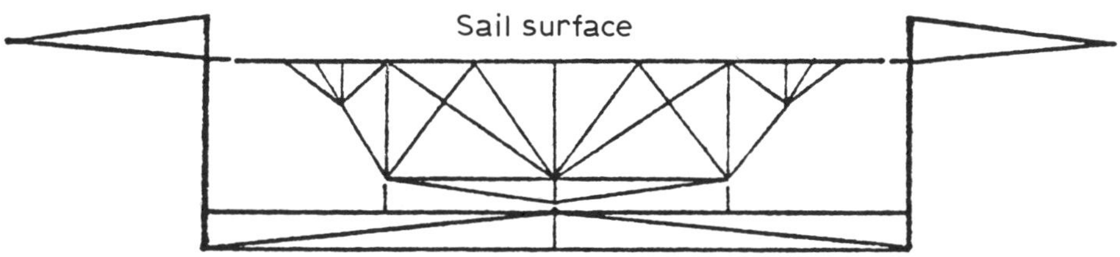

FIGURE 5.19 Cross Section View of Lattice Ship Inside of Scaffolding. (K.E. Drexler)

REFERENCES AND NOTES

1. Jet Propulsion Laboratory (1977) *Solar Sail Technology Readiness Report*, 720-1, Pasadena.
2. Jet Propulsion Laboratory (1976) *Solar Sail Missions and Systems Definition*, 760-154, Pasadena.
3. Jet Propulsion Laboratory (1977).
4. Drexler, K.E. (1977) Design of a High Performance Solar Sail System. *MS Thesis*, Dept. of Aeronautics and Astronautics, Massachusetts Institute of Technology, Boston; (1979) *Design of a High Performance Solar Sail System*, MIT Space Systems Laboratory, Report 5-79, Boston.

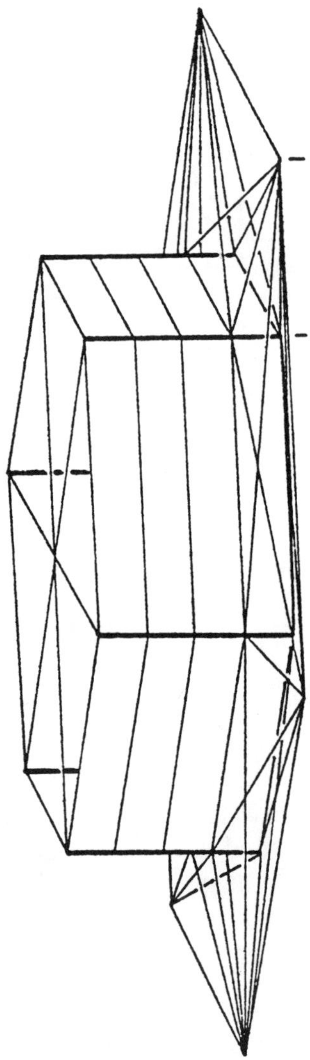

FIGURE 5.20 Scaffolding for Lattice Ship. (K.E. Drexler)

6. SAILING OPERATIONS

The operation of sailing ships includes a wide span of procedures, techniques, logistics, and mission control. The emphasis in this chapter is on the control and maneuvering of sailing ships in planetary and interplanetary space.

The economics of operations is an important factor in the applications of sailing ships to space transportation. Automation and autonomous operation of ships could have substantial benefits in lowering the cost of operations.

ATTITUDE CONTROL OF SAILING SHIPS

Attitude control is provided by changing the center of mass or the center of pressure of a ship. This can be done through the use of vanes, mass movement, sail movement, sail deflection, and possibly reflectivity modulation. The attitude control system can use control responses proportional to the size of the sensed attitude error and the rate of attitude drift. Some form of applied force, such as a gas thruster, can also change the attitude, but these are not desirable for extended sailing voyages or sensitive structures.

Configurations using individual sails for control, where the sails are large relative to the ship, will experience a reaction problem when the sails are moved in some modes. When large sails are moved forward or aft for pitch or yaw control, the ship as a whole will rotate in the opposite sense. Numerical simulations of attitude dynamics should be done to study the controllability of a configuration, especially when these modes are used. Schemes using in-plane translation of the sails to achieve pitch and yaw could avoid this problem.

A guidance and navigation analysis done for the Halley Rendezvous mission showed that an attitude error tolerance of about 0.25° was acceptable. Trajectory deviations caused by attitude errors can be removed on a frequent or even continual basis as the onboard computer compares its measured position and velocity against its desired position and velocity. Alternatively, a less sophisticated approach could be used based on tracking by terrestrial systems, with updated steering commands periodically sent to the ship. In either case, it is not desirable to force the ship back to an original trajectory when errors develop. The best approach is to determine a new optimal trajectory based on the measured position and velocity of the ship.

The control rates must be such as to return the sail to its desired pointing direction with approximately zero residual rate within some characteristic time following a disturbance or drift from the null position. This characteristic time may be an hour or more for a 0.25° correction, or it might be only a few minutes.

The desired pointing directions will change during interplanetary cruise at a generally low rate of less than about 1° per day, with the ship having a few hours in which to reorient itself for those changes. Close to the Sun the pointing direction can be changing at rates near 5° per day, and even higher when changing the inclination of the orbit.

In planetary orbits, the requirements for reorienting the force vector are more substantial. The force vector may need to be reoriented through angles of up to 50° in about 15 minutes in Earth orbit if an optimum steering profile is to be used. Less stringent times or angle changes can be used in a less optimized profile. These reorientation times are generally not a problem with respect to the possibility of the sail flapping loosely during the maneuver because the solar pressure keeps the sail pushed tautly into position.

The attitude control system is required to balance the pressure distribution on the sail relative to the center of mass. This is equivalent to saying that the total force vector must pass through the mass center of the ship if an applied torque is to be avoided. Most of the time spent in cruise conditions will be with the sail force balanced. The conditions needed for balancing the ship at various angles from the Sun cannot be predicted with high accuracy because of knowledge uncertainties in the exact sail shape, structural deflections, and the center of mass. Fortunately, the abilities of the control system can accommodate these initial uncertainties. Within a short time after getting under weigh, the balance conditions can be precisely determined through experimentation.

The *Mariner 10* spacecraft was partially controlled by solar sailing techniques during its voyage to Mercury and between subsequent encounters with the planet. This was done by differentially twisting the solar panels to create torques about the roll axis. During a two week period, roll attitude was precisely held by small torques applied many hours apart. This was done by commands from the mission controllers since there was no autopilot function onboard for that task. This endeavor showed that solar pressure can be successfully used to control the attitude of a spacecraft, and that precise balance conditions can be achieved with experience.

Some ship configurations have attitude singularities, which are orientations relative to the Sun in which the ship has little or no attitude control capability. A ship entering an attitude singularity will almost certainly coast through by its own momentum, so that attitude control can be regained after a few minutes. If not, orbital motion relative to the Sun should eventually provide an attitude in which control will exist.

Gravity gradients in planetary orbits and close solar orbits can cause torques on sails which must be allowed for in either the control system design or in operational procedures. Some sail configurations have the potential for becoming locked into gravity gradient stabilized attitudes. If this were to happen, it may be necessary to maneuver the ship farther out from the central body, if possible, before full control can be regained.

It may be possible to vary the reflectance of a sail or parts of a sail as a means of altering the orbit or attitude of the ship. The ratio of highest to lowest reflectance must be competitive with the effectiveness of reorienting the sail or vanes for the same purpose. The variable reflectance approach also needs to show benefits in the ship's systems, such as mass, power, and complexity advantages.

Control Requirements and Approaches

The full range of operations must be considered in determining the control requirements for a sailing ship. This task is more difficult for general purpose ships that will be used in a variety of assignments.

Turning rates during interplanetary cruise create generally benign control requirements. The more difficult requirements come from planetary spiral trajectories, rendezvous, docking, payload transfer, gravity gradients, and stationkeeping.

The stability of the ship should be kept near neutral. Slightly positive or negative stability is acceptable for nearly all cases.

Analyses and simulations of attitude changes are essential. In particular, gravity gradient effects and orbital motion effects require simulation of both trajectory and attitude motion.

Rigid Sailers

These ships must have full three-axis control during normal cruise and during stationkeeping, when the edge of the sail may be oriented toward the Sun. This means that pitch, yaw, and roll control should exist for all attitudes relative to the Sun.

Pitch and yaw control can be provided by at least four means:

- rotating sail panels
- rotating vanes located at the boom ends
- shifting the sail to move the center of pressure
- shifting a small mass to move the center of mass.

These techniques are illustrated in Figure 6.1 Roll control is done through the use of movable vanes or through rotation of sail panels.

For the first method, where four triangular sails are used, the corners are supported at the vehicle core and the boom tips as described in Chapter 3. The inboard corners of the sail panels can be moved fore and aft (±X) to

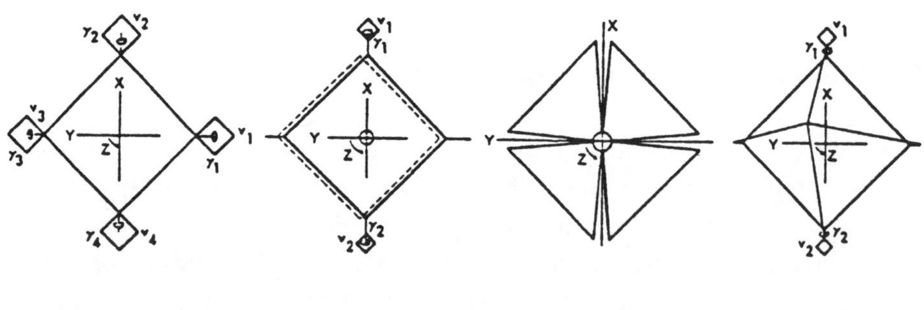

FIGURE 6.1 Methods of Sailing Ship Control. A) Vanes, b) Sail movement, c) Sail panels, d) Ballast mass. (JPL)

provide pitch and yaw torques, which may be the best approach to achieving pitch and yaw control.

At the boom tips, the sail corners may be attached to small spars that can rotate around the boom, providing the ability to raise or lower each outboard corner of a sail panel. The combinations of motion for vehicle control are shown in Figure 6.2. Roll motion should be done using opposing pairs of sails. Optionally, pitch and yaw torques can be achieved by a coordinated movement of the inboard corner and the outboard corners of one or more sails. When the ship is at an angle to the Sun, an opposing pair of sails can be moved forward or aft together to achieve pitch or yaw without causing an adverse attitude reaction by the ship.

Without the tip spars acting as control arms, a similar effect is obtained

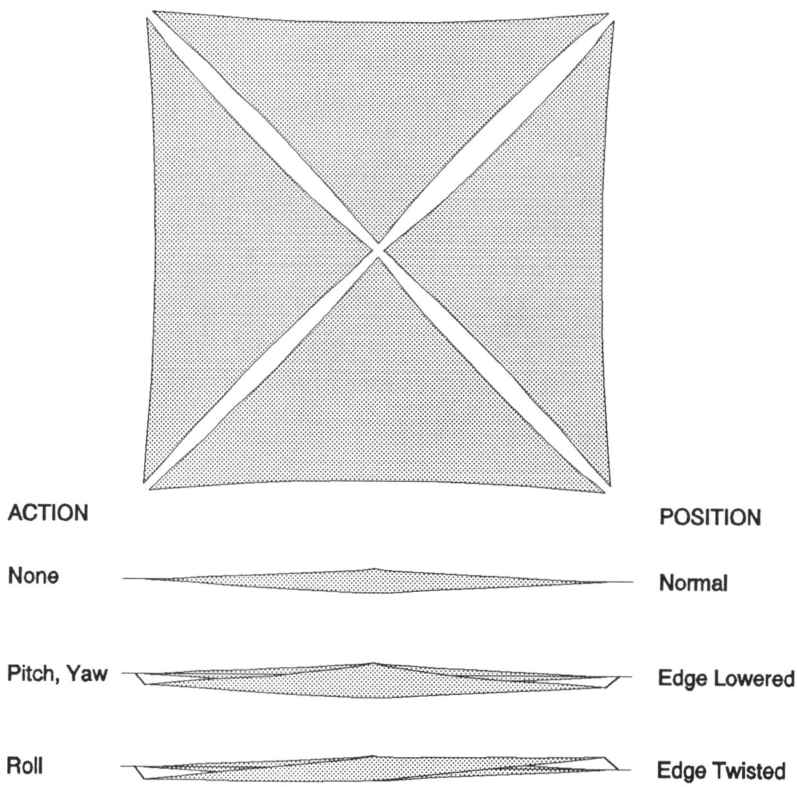

FIGURE 6.2 Attitude Control With Four Sails.

by letting out the lines to the outboard sail corners, allowing the sail panel to billow more fully. If both outboard corner lines are let out together, a pitch or yaw torque can be created. If only one outboard corner is let out, roll torque is achieved, which should be done while giving the opposite sail panel a complementary movement.

The tip spars have the advantage of providing control at all times, including when the sail is edge-on toward the Sun. This control is important for a ship that needs to orbit small bodies or do other stationkeeping without thrust. If a ship does not have these requirements, the simpler approach of using lines alone might be preferred.

In the second control method, two or four vanes can be used at the boom tips, each vane with one or two degrees of freedom, shown in Figure 6.3. Use of a vane at each corner, each with two degrees of freedom, provides the greatest control of the vehicle. This can be important for special maneuvering requirements, such as stationkeeping. Each vane is a small sail in its own right, with its own mast, booms, and stays in various configurations. In addition, external stays may be needed to help carry forces from the vanes to the ship's primary structure. The result is that vanes will be complex structures or relatively massive.

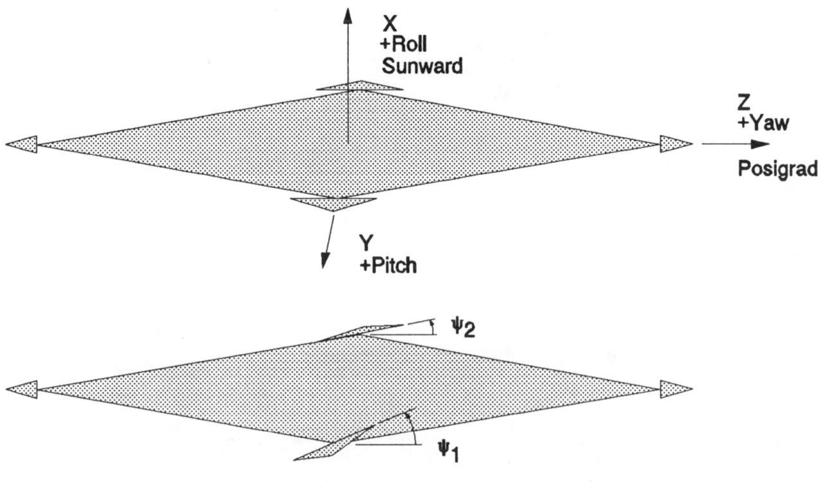

FIGURE 6.3 Attitude Control With Control Vanes.

The vanes can be set at angles independent of each other to obtain combinations of pitch, yaw, and roll. Adequate control can be obtained from vanes with one degree of freedom that rotate about axes parallel to their booms. The nature of these pitch/yaw and roll torques are shown in Figure 6.4. An opposing pair of vanes can be set according to the curves in the figure to obtain pitch or yaw torques without inducing a roll torque. This is done by setting the two vanes to angles having the same roll torque but different pitch torques, so that a net pitch torque exists on the ship.[1]

If yaw movements are only infrequently needed, the ship could be equipped with just two vanes. Yaw change can be achieved by rolling the ship 90°, making a pitch change, then rolling back to the original orientation. A ship equipped with vanes may also be able to use modest translations of the sail as an additional means of controlling the orientation.

In the third method, a solid full-span sail can be shifted by pulling on one of the outboard support lines while relaxing the opposite line. This causes a modest lateral translation of the sail, resulting is a shift in the center of pressure relative to the center of mass. This technique can be used for pitch

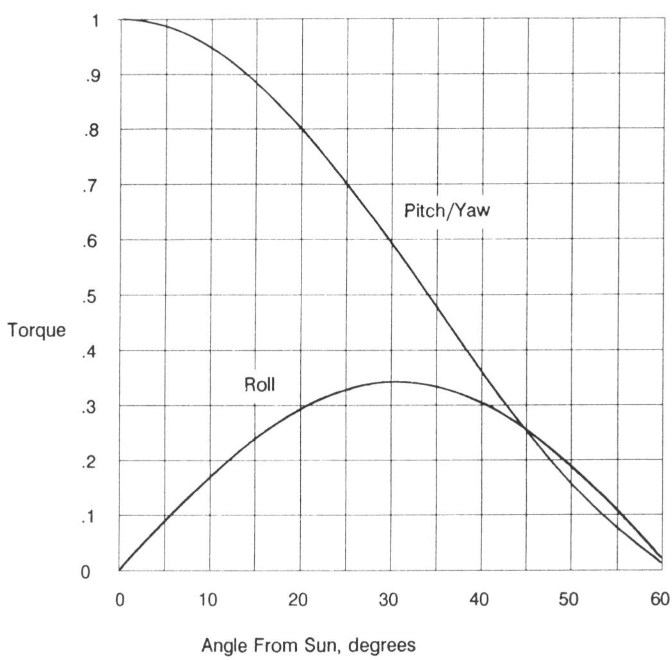

FIGURE 6.4 Pitch, Yaw, and Roll Torques from Rotating Vanes.

and yaw control, but not for roll control. This is an easy technique to implement, and can supplement the other approaches without the requirement for additional hardware. If special lines are run from the core to the centers of the outer edges, the lines can be used to pull in the sail edge to create pitch and yaw torques.

If the ship has four individual sails, one or two opposing sails can be translated to create pitch or yaw torques. Coordinated movement of all three attach lines causes a sail panel to translate. This technique is usable with or without the presence of tip spars.

In the fourth method, the vehicle can use mass translation for shifting the center of mass relative to the center of pressure. A concept that was investigated at JPL is towing an inert mass using four lines that attach to their respective corners of the sail support structure. It was found that the mass can be quite small and still provide substantial design margins for pitch and yaw control. A mass of about 100 kg was considered for the Halley Rendezvous design.

The most attractive approach may be the technique of making modest translations with individual sail panels for pitch and yaw, and rotation of opposing sails for roll control. Using small spars at the boom tips would provide good control for stationkeeping, but using four individual sails without tip spars should also be adequate.

As a supplement for translation control, small gas jets can be mounted on the boom tips to thrust inward for auxiliary motion control. This may be beneficial for docking or stationkeeping around small bodies such as asteroids.

Spinning Sailers

The attitude of spinning sailers is altered by precession of the ship. Attitude control can be provided by several potential means, including:

- use of vanes on the sail outer rim
- rotation of sail panels
- radial translation of sail panels
- a central arm with a vane
- a central arm with a movable mass.

All of these approaches can provide a torque to cause precession of the ship as it rotates. Turning rates are low because of the large angular momentum of the sail. Vanes or rotatable sail panels are needed to alter the rotation rate of the ship.

Vanes can be located at positions on the outer rim, as shown in Figure 6.5. Concentrated masses can be used for attachment lines to react against. The solar pressure on the vanes will cause them to deflect if an attachment line is let out. Control torques can also be obtained by pulling in the outer edge of the vane.

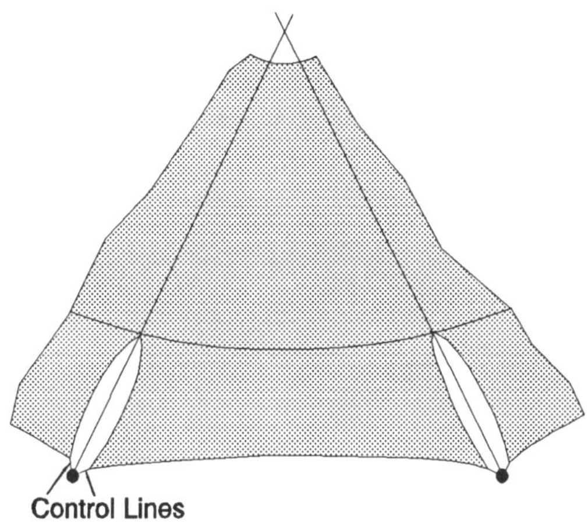

FIGURE 6.5 Attitude Control Vanes for a Disk Sailer.

The rotation of sail panels shown in Figure 6.6 is similar to the rotation of the vanes described in the previous section. The necessary angular deflections of the sail panels are smaller than those needed for the vanes because of their large difference in size. This technique can provide full control of the ship.

The sail panels can be translated radially inward and outward using the attach tendons that hold the inner corners of the sail panels, Figure 6.7. This technique can create large torques for precessing the ship, but cannot alter its rotation rate.

An arm can be mounted on the central body which either has a mounted solar pressure vane or has guides for the translation of a control mass. These provide control torques by center-of-pressure and center-of-mass shifts, respectively.

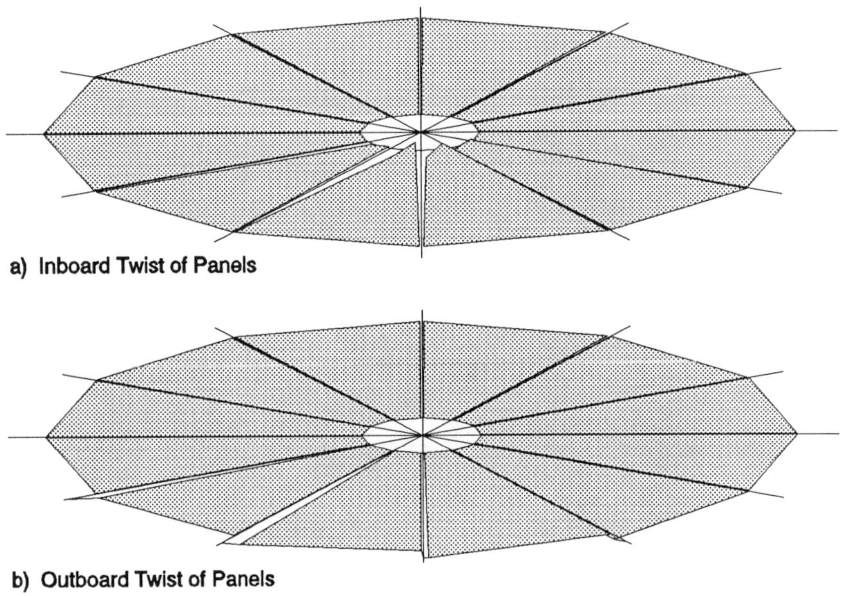

FIGURE 6.6 Rotation of Sails for Attitude Control.

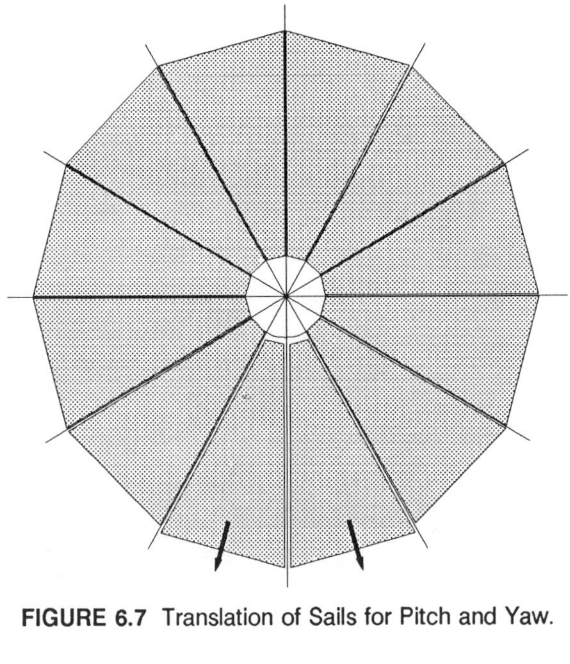

FIGURE 6.7 Translation of Sails for Pitch and Yaw.

A drive motor rotates the arm to the correct orientation to apply the desired control torque. Twisting this vane can apply a torque to change the rotation rate of the sail by reacting against the drive motor. The torque available for rotation is very modest compared to that obtained by rotating panels of the sail.

The coning of the sail under the effects of solar pressure causes an off-center force if the sail does not face the Sun exactly. This will create a continual tendency of the sail to precess unless compensated for by the control system. The control system can use any of the described techniques to control coning precession, but all of them must be operated in a cyclic manner to keep the solar pressure balanced on the sail, except in the case of the vane or mass on a control arm.

Control operation is complicated by the flexibility of the sail and the attachment stays. No torque can be transferred between the sail and the center body without the extended hub that was described in Chapter 3. The control system analysis must take into account the flexing of the sail panels and tendons, flexing of the attachments, the possible existence of deflection waves in the sail, and the transfer of torques between the central body and the attachment stays.

The simplest control approach appears to be through rotation and translation of sail panels. This requires letting out and taking in the attachment lines through some mechanism such as a reel. A vane on a despun arm is also relatively simple to implement, but it does not provide as large a control torque. It also means additional mass and cost.

Where there is a set of sail panels or blades, it can be effective to twist the roots of the sails in either a collective or cyclic manner to alter the pressure distribution across the sail disk. This technique is applicable to the disk sailer, the heliogyro, and possibly other configurations.

Heliogyro

The orientation of the force vector of the heliogyro can be controlled by two techniques. One is by pointing the entire vehicle so that, when the blades are flat in their planes of rotation, the resultant force vector is oriented in the desired direction. The other method is to create a dynamic blade pattern that results in a lateral force component relative to the axis of the ship.

The reorientation of the vehicle is done through twisting the sail blades at their roots, which causes some degree of twisting along the length of the blades. There are two modes of blade twisting called cyclic and collective pitch, illustrated in Figure 6.8.

This cyclic mode of blade twisting can result in a force imbalance across the disk of rotation causing the entire vehicle to precess. These patterns typically require the blade roots to be repeatedly twisted during several rotations of the ship to build up the necessary amount of blade twist.

The other technique of achieving a lateral force component uses repeated cyclic blade twisting to develop a standing blade pattern as shown in Figure 6.9. In this technique, the blades undergo a continual twisting as the vehicle rotates, but at any reference point on the disk of rotation each blade will be at the same angle as it passes the point if the cyclic period equals the vehicle rotational period.

In addition to the cyclic mode of blade twisting, there is also a collective mode in which all of the blades are twisted in the same direction by the same amount. The collective mode of blade twisting provides a torque to alter the spin rate of the ship and to modulate the total force on the ship. If the sail is at an angle with respect to the Sun, which is normally the case, collective pitch causes a precession of the ship.

Details of these dynamical behaviors of the heliogyro are contained in Appendix D.

PLANETARY OPERATIONS

Operations in planetary orbits impose requirements for relatively rapid changes in the orientation of the force vector. There are also other concerns, such as possible exposure of the lee side of the sail to the Sun, back illumination of the sail from the planet, radiation belts, and orbital debris.

Executing spiral trajectories inward or outward around a planet appear to be fully practical operations, albeit possibly tedious in the cases of spirals taking several months to perform.

Typical force vector orientations for an escape spiral are shown in Figure 6.10. The geometry of the velocity vector, the Sun direction, and position relative to the planet determines the orientation of the sail. A specific steering strategy is described in Appendix C.

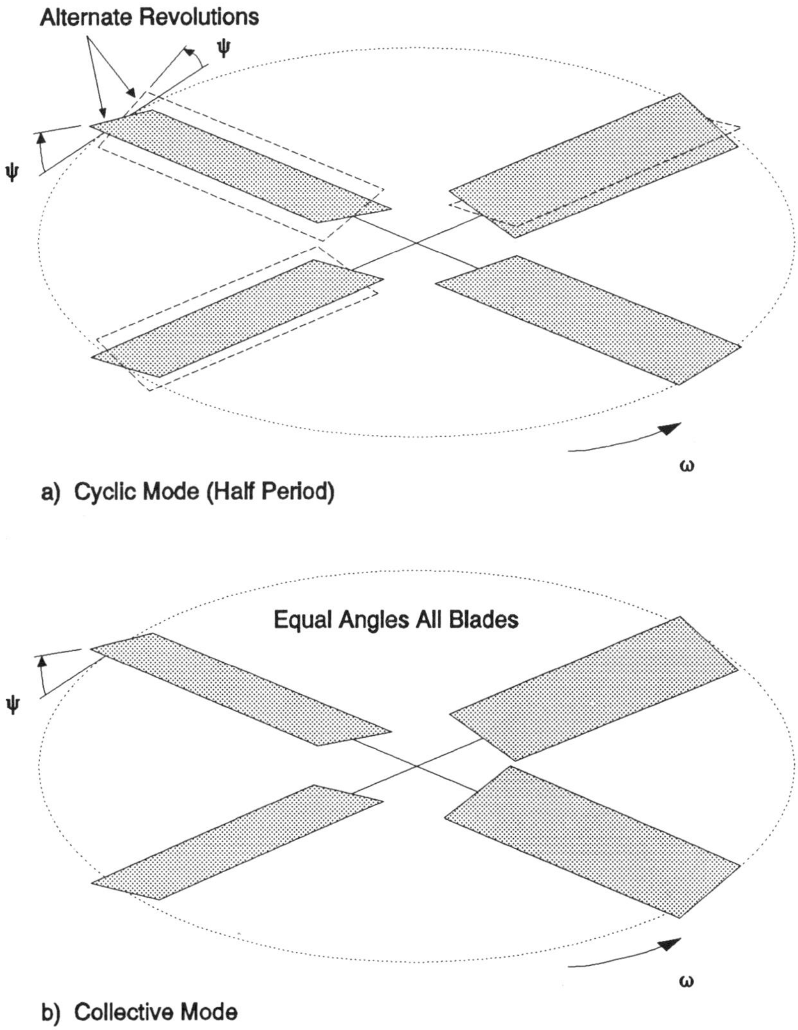

FIGURE 6.8 Cyclic and Collective Pitch of Sail Blades or Panels.

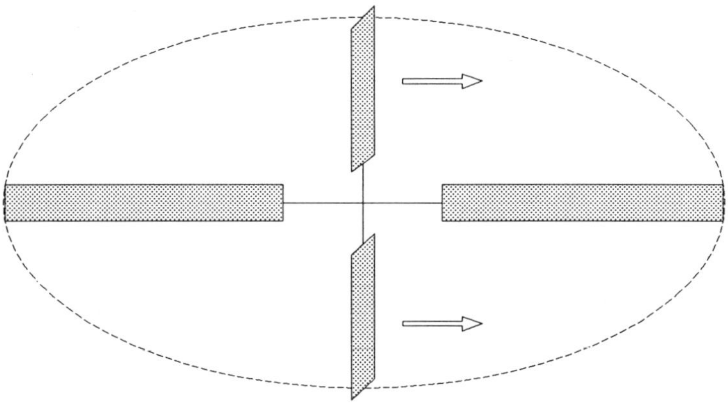

FIGURE 6.9 Lateral Force Components With Heliogyro.

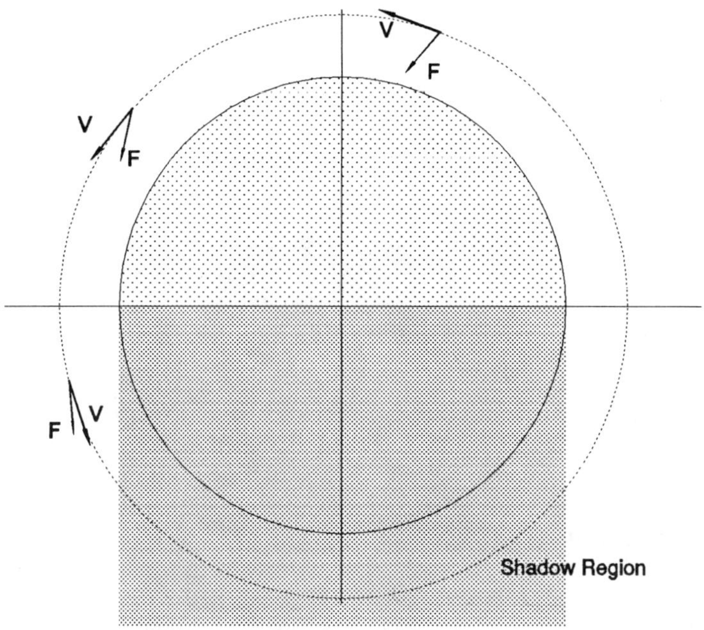

FIGURE 6.10 Typical Steering Vectors for Escape Spiral.

If a spiral trajectory is done in a plane nearly perpendicular to the Sun line, the sail can be set at a constant angle that provides the largest lateral force component. This angle is near 30° for most sail configurations. However, the orbital motion of the central body around the Sun will change the angle between the Sun and the orbit plane, leading to a more complicated steering strategy.

Planetary Escape Maneuvers

The escape maneuver after a long spiral or from a high orbit will typically take from a few days to two weeks. An example escape from an L4 or L5 point is shown in Figure 6.11 for a voyage outward from the Sun. The a_c is 1 mm/s^2, and the escape speed is 1386 m/s at 59.7° relative to a solar rotating frame.

Planetary Capture Maneuvers

Sailing ships entering a posigrade planetary orbit will normally do so from outside the orbit of the planet, so that the ship is approaching the Sun while entering orbit. The encounter is typically on the leading side of the planet in its motion around the Sun. A representative capture maneuver is shown in Figure 6.12 for a ship arriving at Earth with a relative speed of 1600 m/s and an a_c value of 1 mm/s^2.

Simulations have shown that practical arrival speeds at Earth and Mars for a_c values of 1 mm/s^2 are 1600 and 600 m/s, respectively. For a_c values of 1.7 mm/s^2, the respective speeds are about 2000 and 1000 m/s.

Stationkeeping

Stationkeeping with a ship implies holding a particular orbit or position. Holding an orbit requires reducing the solar pressure on the sail to a negligible value, which could be done by furling or relaxing the sails, or by turning the sails edge on toward the Sun. Holding a position or a particular forced, or halo, orbit requires using solar pressure to maintain an equilibrium condition with the body where the stationkeeping is being done.

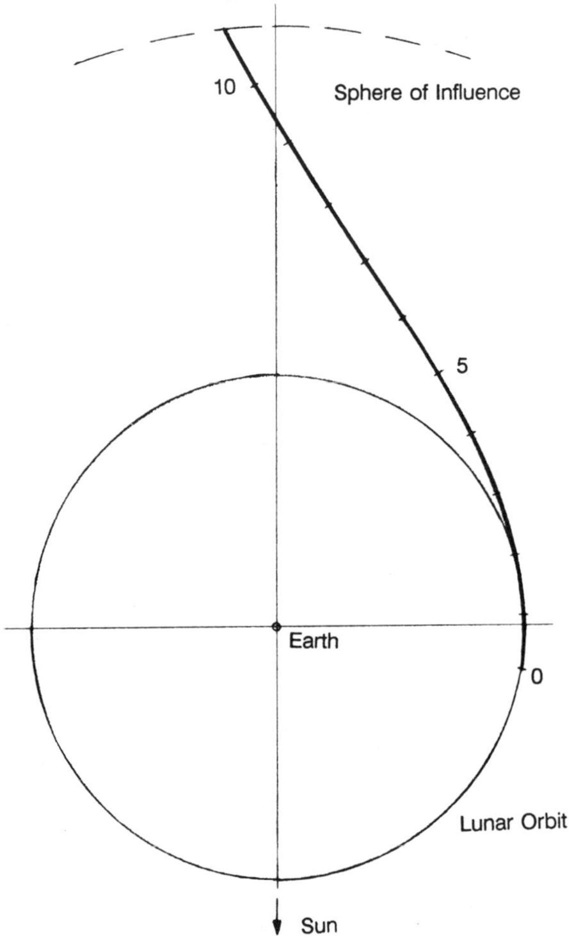

FIGURE 6.11 Escape Trajectory From Earth L4 or L5 Point.

Small Bodies

The transfer from Earth to a small body is typical of other trajectories where a rendezvous is intended. The final approach to the body is from the anti-sun side so the sail can be used for final braking to match solar velocities and insert the ship into orbit around the body.

The ship will probably need to remain in a loose orbit around the body to avoid any need for rapid maneuvering. During this loose orbiting, or

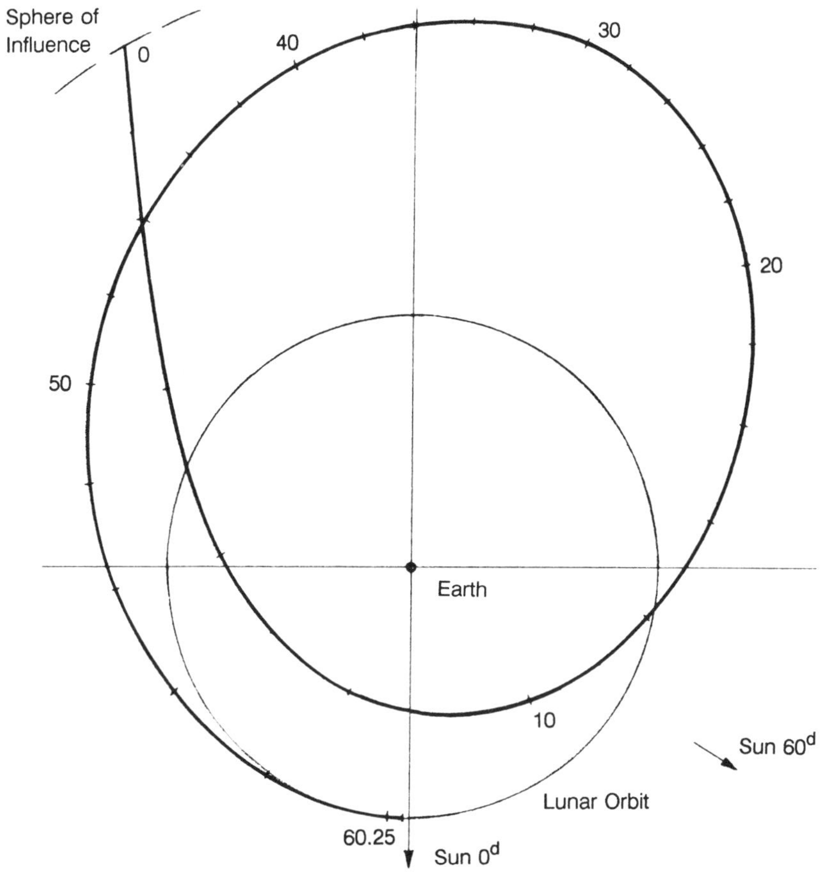

FIGURE 6.12 Multi-Orbit Capture Trajectory at Earth.

stationkeeping, the sail will have to be either kept nearly edge on toward the Sun or reoriented periodically to avoid building up a substantial net velocity away from the body.

Orbits around small bodies are often unstable, particularly when the orbit period has some periodic relationship to the rotational period of a non-spherical body. A spacecraft in such a resonant orbit around an asteroid could impact the body in only a few orbits.

An orbit perpendicular to the solar direction might be successful depending on the size and shape of the body and the characteristics of the sail and ship. The geometry is shown in Figure 6.13. The ship might keep station

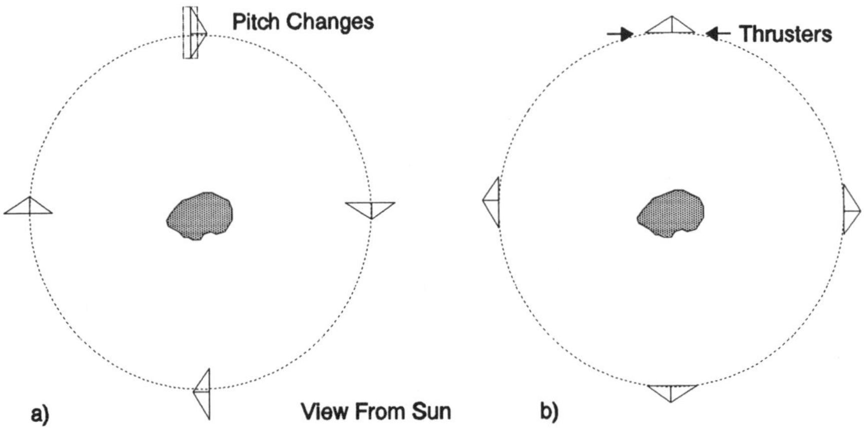

FIGURE 6.13 Example Orbits Around Small Bodies for Sailing Ships. A) orbit control by sail pitch change, b) orbit control by thrusters.

down-Sun in a halo orbit, with the sail kept at a large angle from the Sun. The low solar pressure might be kept in balance with the gravitational pull of the body in this type of orbit.

For all operations around a small body, either entering the shadow region must be avoided or careful procedures must be used to compensate for attitude disturbances. Entering or leaving the shadow means part of the sail is illuminated and part is not. The solar force will no longer be in balance on the ship, creating a tendency for the ship to turn. The low orbital speed means the shadow would move slowly across the sail, which makes the unbalanced condition more difficult to compensate because of the protracted time with the unbalanced forces.

The instability of many orbital geometries may make a small auxiliary chemical propulsion system desirable, for much the same reason that auxiliary engines are used on many maritime sailing ships.

While the ship maintains a remote orbit, a spacecraft can leave the ship to enter an orbit close to the body for detailed mapping of the surface. This spacecraft must have a propulsion system for maneuvering and correcting its orbit, and possibly for landing on the surface. At the completion of its mapping and sample collecting, it would return to the ship, probably leaving instrument packages permanently on the surface.

Tether operations may be practical from the ship for some types of contact, such as obtaining samples from the surface. All of the equipment that will remain with the ship could be brought onboard by a tether to avoid docking operations. The equipment and samples on the tether could be reeled in slowly once the ship is under weigh.

For departure, the sail is turned toward the Sun and in a fraction of an orbit the ship can break free of the body and begin a trajectory to its next destination.

RENDEZVOUS AND DOCKING OPERATIONS

A spaceport is a site where interplanetary ships can berth and where inbound and outbound passengers, crew, and cargo are exchanged. It is also a prime site for making repairs to the ships. A large operation with several ships is likely needed to justify the cost of even a modest station, or dock, dedicated to sailing ships.

The location of the dock must provide a balance of several major factors. The altitude must be sufficiently high to allow solar sailing ships to accelerate without substantial hindrance by atmospheric drag, or the ships must be elevated to those altitudes before being released. The minimum sailing altitude at Earth is around 800 to 1000 km when the Sun is moderately active.

The terrestrial radiation belts affect people, electronics, and some sensitive materials. The dock must be located below about 1000 km or beyond about 20,000 km to remain clear of most of the radiation.

The dock becomes more expensive to reach the farther out it is from Earth. If launch vehicles cannot directly reach the dock, then additional handling operations are needed for the loads being transported to it, which could include the need for an intermediate transfer vehicle operating between the launch vehicle and the dock. Propulsive requirements are obviously greater for the more remote altitudes.

Since solar sailing ships spiral slowly in Earth orbit, there is an advantage from the ship's point of view in having the dock at very remote altitudes. This would reduce the time an interplanetary ship would spend operating in planetary orbit. As altitude increases, a proportionately smaller part of each orbit is spent in Earth's shadow, which helps make higher orbits more effective. The turning rate requirements for the sail also decrease as altitude increases.

The solution to the altitude question is dependent upon the cost of transfers from low orbits, the cost of having a ship spiraling in a planetary orbit, the time value of the load, and other factors. The best solution can be determined only after the total operating scenario is known, but it seems likely that the spaceport would be located above the radiation belts if there is no practical means of elevating the ships to sailing altitudes.

Low Orbit Spaceports

A spaceport in low Earth orbit is suitable for the construction of sailing ships. However, it cannot be used as an operating port because of atmospheric drag affecting the ships unless the ships are moved into higher orbits before being released. One technique is to use very long tethers to elevate the ships to higher altitudes to begin sailing.[2]

A ship with spar-supported sails appears to be the only type suitable for a tethered release from a low orbit station. It might be the only type of ship that can be held by a tether while in a planetary orbit.

When the construction, rigging, and checkout of a sailing ship is completed the ship cannot just simply be released at a low altitude station. Atmospheric drag would cause it to lose altitude in spite of the use of solar pressure on the sail. The ship must be deployed from the zenith side of the station on a long tether so that it starts in a higher orbit. Figure 6.14 shows how tether length alters the apogee and perigee of the initial orbit of the ship.

The tether length may need to be at least 300 km to insure that drag is not a hindrance during an active solar period. This gives a perigee of at least 700 km and an apogee of about 3000 km. The resulting orbit is shown in Figure 6.15, assuming the station is at an altitude of 450 km.

As tether deployment begins, gravity gradient effects pull the ship away from the station, radially outward from Earth. The velocity of the ship is initially too low for the higher altitudes, and the ship lags behind the station (Coriolis effect). The deployment must then be stopped or slowed to allow the gravity gradient to force the ship outward again. As the velocity increases, the tether is payed out some more. The entire deployment is a long operation requiring careful control of the deployment rate to prevent the ship from swinging like a pendulum.

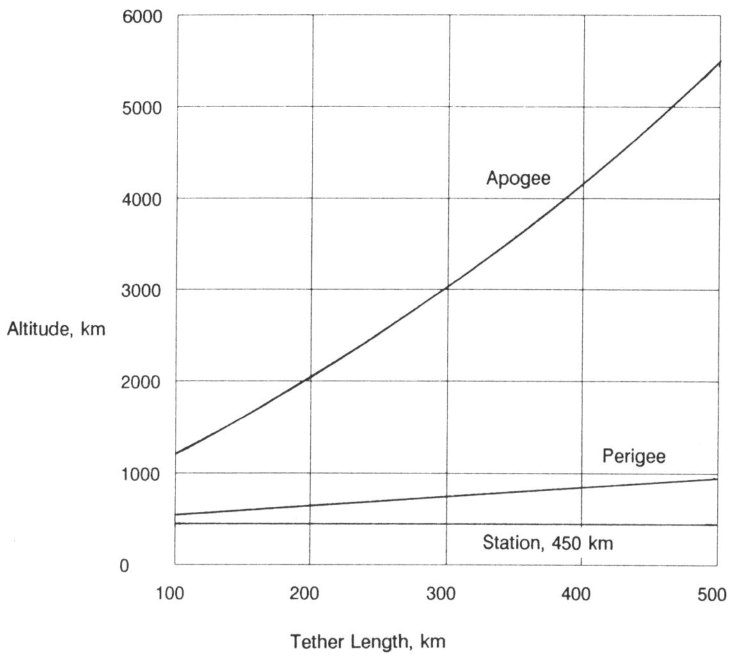

FIGURE 6.14 Initial Orbit Variation With Tether Length.

When the tether is fully out, the ship is released at the point in the orbit where it is moving approximately toward the Sun. This allows raising the perigee first to remove the last concern with atmospheric drag. From this time on, the ship maneuvers through an ever increasing spiral outward from Earth until the it finally breaks free of Earth's gravitational bond.

A fully deployed ship with spar-supported sails can be released at the dock without a tether if a chemical or electrical propulsion unit is used to move it up to a higher orbit. The engines must have low thrust levels and be mounted so that they do not cause damage to the sail or structure. This propulsion system might be an auxiliary unit that is used in other phases of the mission, such as stationkeeping at another body.

A disk sailer or heliogyro could be deployed from a low orbit station if the sails are unfurled after the ship has been placed in a higher orbit by a tether or an auxiliary propulsion system. The sail deployment would then either be

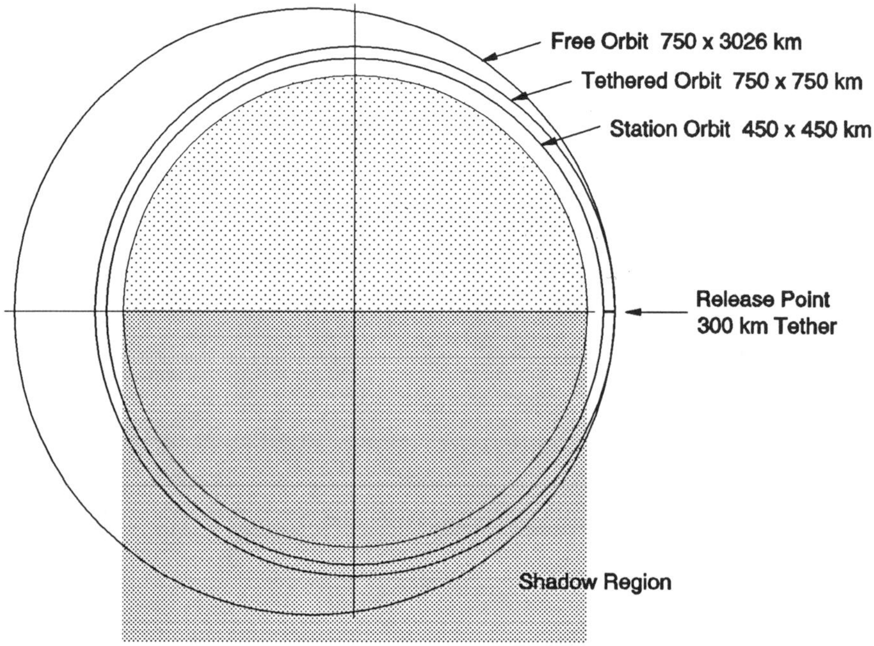

FIGURE 6.15 Initial Orbit After Tether Release.

automated or could rely on a crew that went up with the ship. This deployment crew could return to the station after the ship is under weigh.

High Altitude Spaceports

Ships may simply be met at high altitudes by transfer vehicles operating from a low orbit base. At departure time, the transfer vehicles will return to the base, possibly without leaving anything behind at the rendezvous altitude.

Once the return of interplanetary ships becomes more or less routine, the economics of repeated transfers out to meet the ships will justify leaving some equipment at the rendezvous altitude. This will be the beginning of a permanent dock for interplanetary ships.

At Mars, the moons Phobos and Deimos could be used as spaceports. A ship departing Deimos is shown in Figure 6.16, shortly after separating from its deployment tether.

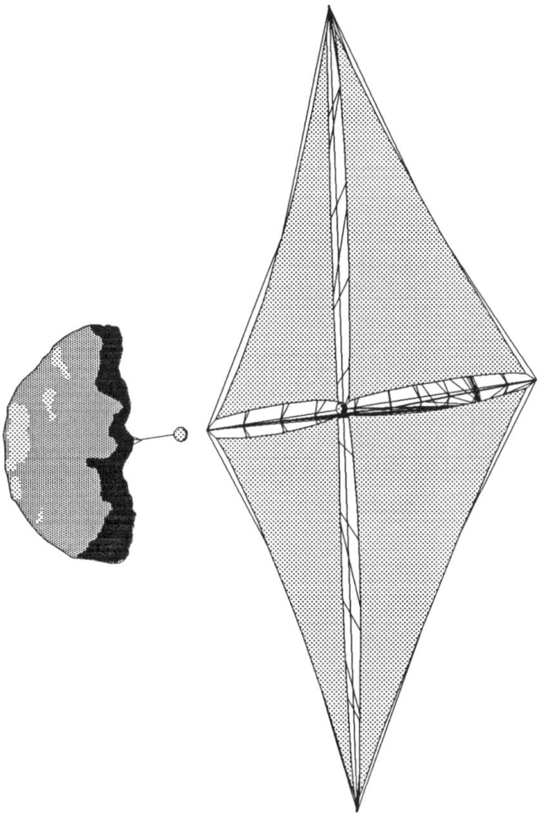

FIGURE 6.16 Departure from Deimos.

Cargo transfers could be done at the L4 and L5 points for specialized interplanetary ships such as the lattice frame ship, which is difficult to maneuver in lower planetary orbits.

Port Operations

Preparation for the arrival of a ship would begin long before its scheduled arrival into port, but the difficult operations begin as the ship makes its final approach. The size and fragile nature of the sail complicate the operations. Some types of ships, such as those with spar-supported sails, can sail in to a facility in low orbit.

The sail will be oriented to remove the remaining relative velocity as the ship reaches the defined rendezvous point. It is unlikely that this point would be at the dock. Instead, it might be a few hundred meters or a few kilometers from the dock. A tether with a maneuvering unit at its end may be sent out and attached to the ship at that time.

A three-axis stabilized ship could be pulled in and firmly secured to the dock. A ship with a spinning sail might be brought all the way in or it might be left on the tether to remain at a large separation from the dock. Some spinning sail designs might not be suitable for tether attachment because of structural or control problems, requiring that the ship remain active to stay in an orbit close to the dock. If a sail cannot be brought in to the dock, then either the ship or part of the ship will be detached and brought in to the dock or a transfer vehicle will be used to shuttle between the ship and dock.

A ship attached to the dock can be fully or partly deactivated. The dock will provide attitude stabilization and power for the ship during this period. A tethered ship will need to remain more self sufficient during the period of attachment.

The dock will provide means of transferring cargo to and from the ship. Servicing and repair of the ship will also be done, followed by a thorough checkout before departure.

If the ship is manned, then the crew will board before departure to aid in the checkout process. Any passengers will board just before departure.

The departure is very easy compared with approach and docking. The ship will be released from the dock or tether at the point in an orbit where it can begin gaining energy. The gravity gradient effect and solar pressure will cause the ship to move clear of the dock or tether without any collision hazard. The release of the ship should be done without imparting sharp forces to the ship.

The spaceport must also have docking provisions for the transfer vehicle coming up from lower altitudes. While the transfer vehicle is attached, the station will provide electrical power and other services to it. The docking provisions will allow for transfer of people and cargo.

The spaceport will need to provide some storage capabilities for the cargoes and quarters for service people, ships' crews, and passengers. It will generate power, and have a life support system when it is in full operation. Similar stations may exist someday in orbit around Mars and Mercury.

INTERPLANETARY OPERATIONS

The economic advantages of sailing ships could lead to extensive interplanetary shipping operations in the coming decades. If this comes about, large numbers of ships will be carrying cargo and people to support the human expansion beyond Earth.

Several sailing ships may be employed in supporting the first manned expedition to Mars. If a permanent presence is then established on Mars, we could see tens of ships operating between Earth and Mars, plus doing other exploration and delivery tasks.

Sailing ships have a broad flexibility in their departure and arrival times, but the trip times are sensitive to deviations from the optimal departures and arrivals. The degree of practical flexibility depends strongly on the acceleration capability of the ship.

Before the middle of the Twenty-First Century, each optimal departure time to Mars could be an occasion for a fleet of five, ten, or more ships breaking out of Earth orbit with people and cargo bound for the red planet.

Robotic Ships

The use of robotic ships can provide important economic advantages. The life support system needed for a crew, including the pressurized quarters, is a major payload penalty. It may turn out that ships will not have permanent crews, and the crew quarters and life support system will be detachable from the ship. Crews may be used only when passengers are being carried.

Robotic ships will operate under some level of remote supervision to assure that the automated actions are correct, and to provide destination and sailing instructions.

In full-scale operations, the control center on Earth should optimize the operation of its fleet of ships to meet the full set of cargo delivery requirements, taking into account the continuous motion of the planets and the ships. It must allow for the acceleration capability of each ship with every load that it will carry. The control center sends destinations and schedules to each ship, and also updates operational data stored onboard.

The operation of a ship can be relatively autonomous. The ship can determine its own trajectories and steering profiles, determining and maintaining the best sail angle to reach its destination. However, it needs to optimize its course taking subsequent operations into account. For example, if a ship has been instructed to deliver a load to Mars and then quickly return empty for another mission from Earth to Venus, it will follow a different course to Mars than if it is to carry part of its load to Mars and the remainder directly to Venus.

En route, the ship will make continuous navigation checks on its progress. It will make guidance corrections to compensate for errors or to take into account better estimates of the time and location of its rendezvous with its destination. This guidance and navigation information will be relayed to the control center which will make periodic verifications that the ship is navigating correctly.

The onboard computer receives high-level directives from the control center on Earth, but most detailed commands are generated onboard with some level of monitoring done from Earth. The ship is self-monitoring, and reports any significant problems or anomalies to the control center. The onboard configuration controller isolates faulty components and selects redundant ones to continue functioning as part of its autonomous capability. The control center will assume more direct control of a ship experiencing substantial difficulties.

The mission control center will have more resemblance to the scheduling center of an airline or shipping company than the government space mission control centers using hundreds of people to support the operation of a single craft in space. This is aided by the autonomy features of the ships, which are especially important in keeping costs low. A technical staff does routine monitoring and problem resolution for the ships.

Towing Operations

Payloads can be towed by ships with a single line. The line should attach to the center of the ships without bridles so that the ships can change their attitude and thereby change the towing direction.

The agile ships with fully articulated spars are good choices for towing tasks. If the towing operations are to be for long durations, as is likely to be the case, robotic ships are probably necessary choices. A towing team of these ships can be used to transport large loads.[3] A team is shown in Figure 6.17 towing a comet fragment.

A 1 km ship can reasonably carry a burden of 30 to 40 tons, so a team of such ships could tow burdens of a few hundred tons. Trip times from near-Earth asteroids and comets to Earth would typically range from 2 to 5 years.

Large, multi-hundred ton structures such as habitats and mining plants could be towed from Earth to the other planets and small bodies of the solar system, similar to the way that drilling and oil production rigs are towed across oceans by teams of tug boats.

Economics of Operations

The economics of using sailing ships is easy to describe, although specific scenarios must be defined for a quantitative analysis. When chemical propulsion is used to send something to Mars, 80% of the mass in Earth orbit is the propellant required just to send the shipment to its rendezvous with Mars. Propellant for stopping, and that for returning to Earth, must come out of the net payload leaving Earth.

Since sailing ships can carry loads to Mars and back to Earth without the expenditure of propellant, excluding that for descending to and leaving the surface, the launch costs to Earth orbit are reduced by more than an order of magnitude. The reduction of hardware necessary for reaching Mars is a further savings. The fact that the sailing ships can be used for many voyages instead of being discarded after a single use is a further economic advantage.

It is too early to know if entire expeditions can be done with sailing ships, but it is clear that the ships should have a major role in supporting expeditions. Since no other form of transportation available in the relatively near term is fully reusable and operates without propellant, it is clear than sailing ships offer an important economic advantage. This being the case, it

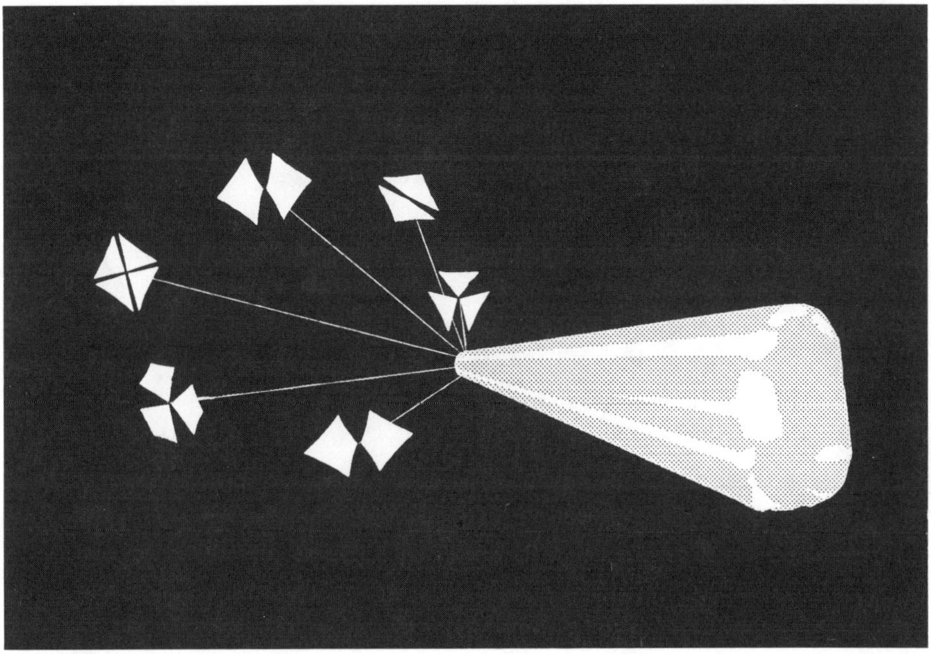

FIGURE 6.17 Team Towing. A team of agile robotic ships has a bagged comet in tow to Earth orbit for research and resource usage.

is quite possible that ships will be owned and operated by private shipping companies supporting the development of interplanetary space.

INTERSTELLAR AND OTHER HIGH-ENERGY OPERATIONS

Solar sailing operations can be used to reach the outer solar system in relatively short times, to reach beyond the region of the planets, and even to travel to other stars. Gaining the necessary energy requires making a close passage by the Sun.

The approach to the Sun is done by setting the sail against the velocity vector at an appropriate point in the orbit. The ship will then spiral in toward the Sun directly, or in multiple orbits, to achieve the greatest possible pressure on the sail. The limits on what is achievable is determined by the thermal tolerance of the sail, the thermal tolerance of the payload and sail support structure, and the mechanical strength of the sail and structure. The thermal limit of the sail is likely to be the deciding factor.

Once the desired perihelion distance and location for the trajectory has been achieved, then the sail is set edge on to Sun so the ship will coast to the point where it can begin to accelerate. At that point, the sail is set to achieve the maximum energy gain. This sail angle may be constrained by the sail temperature limit.

If solar sailing techniques were used to depart a solar system, those techniques can also be used to stop at the destination, provided the destination star is similar to the first, and that the sail has held up adequately.

REFERENCES AND NOTES

1. This approach to using single degree of freedom control vanes was invented by William Layman, Jet Propulsion Laboratory, Pasadena, CA.
2. The tethered deployment concept was invented by the author, and later independently by John M. Garvey, McDonnell Douglas Space Systems Company, Huntington Beach, CA.
3. The team towing concept was proposed by Chauncey Uphoff, 1976, then at the Jet Propulsion Laboratory, Pasadena, CA.

7. BEAMED POWER SYSTEMS

Power beams from lasers and microwave devices can be used to send high energy densities far out into space, even to our neighboring stars. These beams of electromagnetic radiation can be used to enhance the performance of sailing ships by a few orders of magnitude over solar sailing techniques. Travel time to Alpha Centauri, for example, could be reduced from roughly 10,000 years to 30 to 100 years by the use of beams. The power beams can be used for both interplanetary and interstellar ships.

The costs and technical difficulties are expected to be immense for interstellar voyages. The power level and the scale of the structures needed for interstellar power beams are far beyond anything humans have yet attempted. However, if these things are built, it will be with construction and transportation technologies more advanced than what we are familiar with on Earth. Massive robotic assistance in construction is likely to be a major benefit in any such endeavor. The available budget and technology will determine the size, and hence the power level, of the system. The combination of payload mass and travel time will be a function of that system power level.

THE PRINCIPLES OF BEAMED POWER

Light and microwave energy are both electromagnetic radiation, with different wavelengths, that can be used in long range power beams. Both types of power systems use an active solar energy conversion unit to generate the light or microwave radiation, and a lens to create a high quality beam.

The exhaust velocity of light or microwave propulsion is not optimum for interstellar travel. It is easy to show that the shortest trip time for a given exhaust stream power occurs at far lower exhaust velocities. As a result, effective interstellar travel with photonic exhaust requires extremely high power levels, precluding the use of onboard generators and photonic engines. This problem can be solved by collecting solar power and projecting it in a precisely focused beam to the interstellar ship.

The diameter of the Sun, about 1400 Mm, limits the focusing or collimating ability of a reflector or lens system using sunlight directly to create a beam. The divergence is so severe that over a few light years, the beam spreads to a diameter greater than 20 AU. To maintain a useful power density to Alpha Centauri, the total beam power would have to be greater than the output of the Sun. Using reflectors or lenses to project sunlight from the inner solar system to the outer regions is also impractical because of the divergence.

Achieving a beam of sufficiently high quality depends on collecting the solar energy and transforming it by a laser or microwave amplifier into a beam that can be focused by a lens or reflector. Light leaving the laser, or microwave energy leaving its radiating antenna, does not have sufficient quality to project over the distances to the nearest stellar neighbors without serious dissipation. Unless a large, remote lens is used to reduce the beam divergence, a very large laser emitter is necessary to maintain an adequate power density over the full operating range.

Beam Characteristics

Laser and maser systems will create high quality beams. The radiation in the beams is coherent, having a single wavelength, and in phase. A system consisting of a laser and a remote lens can project a focused beam over tens of lightyears, which could enable interstellar travel throughout our immediate galactic neighborhood.

For maximum propulsion effectiveness, the beam size at the sail should be no larger than the sail to minimize the energy output requirements. The beam should be no smaller than the sail to avoid high energy densities on the sail, and to provide a nearly uniform load distribution across the sail. However, a safety margin must be included in the beam width to allow the ship time to recover from drifting away from the beam centerline, and to allow for the lens drifting off its nominal position and attitude. The size of the safety margin depends on the maneuverability of the ship, the ability of the ship to locate the beam center, and the nature of the causes of the deviations.

The lens can focus to a beam diameter B given by the standard criteria for focusing limits, which is

$$B = 2.44\lambda s/D \tag{6.1}$$

for wavelength λ, distance s from the lens, and lens diameter D. This is taken as a practical operating limit, with 84% of the beam energy inside of the given diameter. The 16% of the energy beyond the diameter B is dispersed, so that a significant gradient will exist near and beyond B.

Since the minimum beam size B is proportional to distance s, the power intensity clearly decreases by s^2 for the minimum beam focus. However, by building the system components on a grand scale, the lower limit of the beam size can be made approximately equal to the sail size in the vicinity of the destination. Then by continually refocusing during the voyage, the maximum available power can be maintained on the sail.

Some of the system design requirements immediately result from this approach of keeping the full beam power on the sail. A variable focus lens must be used, and made large enough relative to the sail so that the diffraction limited focus near the destination is approximately the size of the sail. The distance of the lens from the emitter, the beam spread from the emitter, and the emitter diameter must be matched to make maximum use of the emitted power.

An alternative is to use a fixed-focus or limited-focus lens in a system of greater output power than necessary with a focusing lens. A fixed lens might get only a small fraction of the beam power onto the sail at the beginning of the voyage. This is a size versus complexity cost trade that can be made only when reliable cost data is available.

A safety factor of 40% on lateral dimensions is assumed to be reasonable. This means that the ship can drift off course no more than 40% of the sail diameter, even in the event of a control system failure. This 40% factor means that the beam power must be twice the power needed on the sail.

The beams should not contain any significant outward radial gradient across a sail or in its immediate vicinity. Such a gradient would create a turning torque on the ship, causing a change in attitude which could result in the ship being ejected from the beam. Whether or not this condition can be met, or if a beneficial gradient could be created in the beam, can be determined when specific engineering studies are done.

A velocity v can be achieved in straight line acceleration over a distance s by the lens maintaining a constant power level on the sail, given by

$$v^2 = \frac{4sP\eta}{cM} = \frac{4sW\eta}{c\sigma} \tag{6.2}$$

for spacecraft mass M, light speed c, power level P, power density W, areal density σ, and efficiency η.

The energy levels in the output beam are in the hundreds or thousands of terawatts (TW), or 10^{12} watts, for manned ships, which contrasts with current planet-wide power consumption of slightly over 1 TW.

Ship Concepts

For interplanetary operations using beamed power, the ships can be ordinary solar sailing ships. There is no need to make any design changes for using beam power, but ships meant to primarily operate in beams might have design parameters such as allowable illumination intensity selected for best advantage.

The sails used by the interstellar ships should have a low areal density, high temperature tolerance, high reflectivity, and high emissivity. The sail supporting structure should have high strength and resistance to dust erosion.

The ships must be designed to support the people onboard on an almost indefinite basis with reasonable comfort, but with as little mass as practical within design constraints and safety.

A simple sailer can be used as an automated probe, while one carrying people needs a means of stopping. This can be done with a mixed mode propulsion, using some type of rocket engines, or by using a two-stage design. A three-stage design could possibly be used in making a return trip.

The ideal approach for mechanical simplicity, and possibly lowest cost, might be the use of a high-drag sail. It will accelerate normally when illuminated by a power beam, taking the ship through the first part of its voyage. In the deceleration phase, the sail is used to create drag forces to slow the ship. This can be done using the power beam at low intensity as a source of electrical power to drive large coils that collect charged particles over a large volume, creating a resistance to the motion. The coils can be structural elements in the sail, and might be able to work as superconductors because of the low background temperature of space. The sail itself might also be able to collect interstellar dust to add to the drag force. If so, then the sail would be kept edge on to its motion until ready to decelerate. The terminal phase of the voyage would use the destination star for final deceleration.

Interstellar Voyages

Trip times will be long even with beamed power systems, ranging from a few decades to a century or longer. The planning for survival after arriving at the destination will need to include the likelihood of having to live in planetoids constructed by the settlers and their robots.

The voyages might be made using ships that can decelerate themselves, or using a decelerator beam or reflector at the destination. The decelerator device could be set up by the first settlers, constructed by robots sent ahead of the first manned ship, or it might be the second sail of the first two-stage ship to reach the destination.

The strategy for supporting the first people arriving at the destination may likely require sending additional ships more frequently than one every transit period, so that several ships could be in transit at the same time. Ships must avoid shadowing one another, so they must travel in modes that do not cause interference. Some of the possible modes include:

- using assigned sectors of a single wide beam
- using small beams dedicated to each ship
- time sharing a single beam
- using large separations in a single beam.

Time sharing is based on using high power levels for relatively brief periods, perhaps 1 to 10 years. In this mode, one ship would be accelerated to coasting speed, and it would then move away from the beam center. Another ship would then enter the beam for acceleration. The deceleration phase would be done in a similar manner. Ships not undergoing acceleration might carefully maneuver into the first null ring around the beam center, allowing the beam power to reach a more remote ship. The ships would arrive separated in time by whatever the cycle time is for sharing the beam.

The beam could be shared by several ships without any having to leave the beam center if the separation distance, beam diameter, and beam power were adequate to get the necessary power density on the more remote sails. This approach relies on the beam spreading, or filling in, that would occur forward of each ship. If this approach is practical, the time separation between the ships might be quite small.

INTERPLANETARY LASER SYSTEMS

The laser conversion units operate on concentrated solar power that is converted into coherent, collimated beams. These devices are organizers of the electromagnetic radiation which they receive. The sunlight collected as the driving power arrives in nonparallel directions, with mixed wavelengths and phases. The photons are emitted nearly parallel, with the same wavelength and in phase. A lens improves the quality of the beam, but it may not be needed for interplanetary applications, depending on sail size and distance as well as parameters of the laser emitter.

The high orbital rates of the ships and lasers create tracking problems for laser systems located in the innermost regions of the solar system. Synchronization of a lens separate from the laser could be a serious problem.

The most effective approach in reducing travel times is to place lasers at all destinations having high traffic densities. Starting and stopping ships at planets, small bodies, and other sites by lasers at those locations yields high acceleration levels in generally optimal orientations for minimum time voyages. Trips between Earth and Mars could be reduced to about a month instead of the 3 to 6 months required for ordinary solar sailing ships.

The best location in terrestrial space is either on the Moon at one of its poles, or in a high Earth orbit. Beam steering on the lunar surface could be done with a movable mirror.

Sailing throughout the solar system would be substantially improved even with only a single laser, that could be located in terrestrial space. This would be the best location for the first laser. This single laser would not only be used for departures and arrivals at Earth, but the beam could be directed to ships operating in interplanetary space or in orbits around other planets.

When a second laser is built, it could be placed at Mars to minimize travel time between Earth and Mars, in what is almost certain to be the region of greatest shipping traffic. An alternative is to place a reflector at Mars to improve the effectiveness of a terrestrial beam over using the beam directly for arrivals and departures at Mars.

Using a terrestrial and a martian laser system for driving sailing ships gives travel times that are about the fastest possible of any propulsion method, without the cost and hazard of nuclear reactors, and without the costs of propellants and engines.

INTERSTELLAR LASER SYSTEMS

The scale of a laser system for interstellar operations requires a large lens remotely located from the laser emitter or a laser emitter made large enough to compensate for not having a remote lens. The optical quality of the system need not be perfect, since imperfections can be compensated for by increased power. This allows effective use of the zone plate lens for creating a collimated beam that can span across tens of lightyears of space before undergoing substantial dispersion. The concept and example data are from Robert Forward.[1]

Laser Design

Laser technology is changing so rapidly that the best type of laser that could be manufactured in space on an immense scale many decades in the future remains an unknown. There are several candidate lasers available with current technology,[2] so we know that something can be found that will work, but at some unknown cost.

We do know that the laser must have a large diameter, essentially completely filled. The emitter will almost certainly be made up of many individual lasers that must work together coherently to achieve the necessary beam quality. This has been demonstrated in the space weapons program. A long operating life is desirable for the individual lasers, but not essential if they can be replaced during operations.

Although the total power output of the laser bank must be high, the output of the individual lasers is not necessarily extraordinary. The daunting problems are making, operating, and coordinating the large number of lasers needed to fill the laser array, and achieving an operating lifetime for the lasers that leaves the project an economically viable endeavor.

A laser emitter may operate at a wavelength of about 1 μm. This value is assumed in the following calculations.

Laser Construction

The laser conversion units could be located in solar orbit, in mercurian orbit, or on the mercurian surface.

The material for their construction could come from Mercury. Since it is a small, airless planet, slightly larger than the Moon, electromagnetic launchers could be used to move the materials from the mercurian surface into solar orbit. Sailing ships could be used to collect the materials after launching, and deliver them to their intended site.

Minor planets, such as Icarus, could also be used if they contain the appropriate elements. The mining and processing would be done under essentially weightless conditions at these small bodies. Use of the minor planets would be more difficult with respect to the orbital mechanics involved because they have orbits which extend out into the solar system far beyond Mercury. Solar sailers would be needed to maneuver the materials from the bodies to the construction site.

Alternatively, Icarus or other small body could be both a source of material and a massive foundation around which the laser could be constructed.

The units in solar orbit do not need to be made as a low-mass design since they remain in captive orbits, although the areal mass density will in part determine the orbit of the unit. However, it appears to be most advantageous for the emitter to remain in a fixed position in the solar system.

Laser Operation

Unless the laser system is located on Mercury, the orbital path will be displaced from those of unperturbed bodies because of the reaction from the momentum of the beam. The orbital plane should be perpendicular to the direction of the beam. For an interstellar application, the beam direction remains essentially fixed, so the orbit does not change appreciably. The orbital motion of the laser in a solar orbit, or on Mercury, imposes an orbital motion and variable pointing requirement on the remote lens to maintain the beam on the sail. The laser emitter might also be placed in a precessing polar orbit around Mercury, with its output sent to a deflector at the Sun-Mercury L2 point, then to a lens and the ship.

If the areal density of the laser system is held to approximately 1.4 g/m^2, the acceleration of the system is equal to the solar gravity, so the laser system need not experience any orbital motion, as shown in Figure 7.1. The laser beam would then have a constant direction throughout the voyage. This will simplify the pointing of the emitter and the positioning and pointing of the lens.

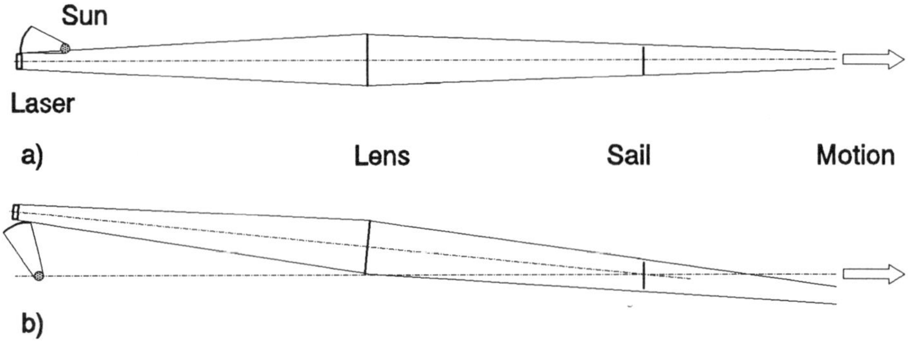

FIGURE 7.1 Laser and Lens Geometry. A) Laser and lens stationary on sailing line, b) laser and lens in halo orbits around sailing line.

If it is not practical to have this low areal density, the next best choice might be to place the laser station in a tight circular orbit around the Sun that is perpendicular to the beam line. The solar pressure and beam pressure on the laser will cause this to be a halo orbit as shown in the figure.

The laser output could be corrected based on a pilot beam sent back from the ship. This beam will pass through the optical system in a reverse direction, and provide information on how to correct the emitted wavefront. This correction would be done by adjusting the coordinated operation of the numerous lasers in the system.

Lens Design

The size of the lens, at 1000 km, necessitates making it with a low mass and a simple design. This can be done by making a lens of the Fresnel type. A lens of reasonable optical quality can be made by using thin, flat rings that alternate with open spaces, as shown in Figure 7.2. The thin rings slow the light passing through enough to shift the phase by one-half of a wavelength relative to the light passing between the rings. The wavefront that emerges from the lens can be a good approximation of a smooth, converging wavefront that has a definite focal point.

The focal point is set ahead of the sail so that the beam fully illuminates the sail. The focal point distance is a function of the ring spacing and the wavelength, defined by the expression

$$f^2 + r_n^2 = (f + \tfrac{1}{2}n\lambda)^2 \tag{6.3}$$

for the focal length f, the radius r_n of the nth ring, and the wavelength λ. The example lens has a total of 111,410 rings. The outermost ring has a width of 2.25 m, and is the smallest ring.

The zone plates can be made of flat plastic sheets, with a thickness of about 1 μm. The performance of the lens could be improved if a taper was made in the plates in the radial direction. The mass of this lens with flat rings is approximately 560,000 metric tons.

The operating life of the plastic is an issue that must be thoroughly evaluated. The lifetime is a function of the intensity and wavelength of the illumination. A large area ratio between the lens and sail would lower the intensity on the lens, which would help with the lifetime. Selecting a wavelength, or range of wavelengths, to which the plastic is relatively insensitive should also help with the lifetime.

The ring tolerance to geometrical errors is 0.56 m radial average and 5 Mm along the axis. The radial tolerance is demanding given the 1 Mm diameter of the lens.

The variable focus requirement might be met by changing the Δr spacing of the rings. Implementing a variable ring spacing would be difficult because of the complexity due to the large number of rings and the mechanical precision required. The outermost rings must be accurately positioned to about one part in one million.

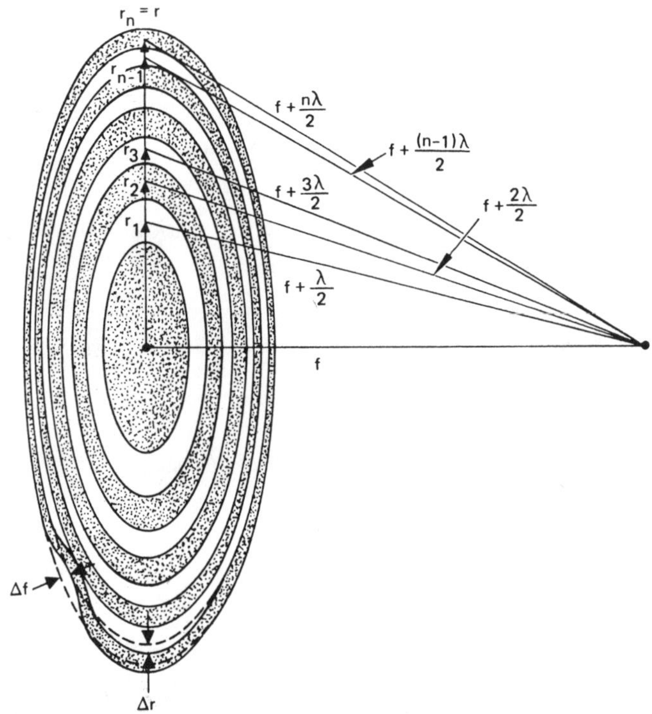

FIGURE 7.2 The Fresnel Zone Lens. The focal distance is a function of wavelength and ring spacing. (R.L. Forward)

One possible approach is to use a large number of butterflies around the outer edge in conjunction with tuned tension springs between the rings. Changing the angles of the butterflies will alter their pull on the lens that might successfully change the ring spacing in the necessary proportions. A spinning lens might successfully use adjustable masses at the rim, but this may be complicated by the dynamic Coriolis effect.

Increasing the Δr spacing would likely leave gaps in the rings, which would degrade the effectiveness of the lens. In addition, when the ring spacing is changed, the ring widths should also be changed to maintain full effectiveness of the lens.

Wavelength changes in the light could potentially be made if free electron lasers are used. These lasers are able to operate over a wide range of wavelengths.

Lens Construction

The size of the lens makes the use of an external construction frame undesirable, or even impractical. As a result, the lens should be constructed while spinning or while held under tension by other means.

The ring tolerance is a problem if the lens is spinning, due to centripetal load effects causing stretching of the structure. The extent of stretching during construction changes due to the increasing mass of the structure. Working with this effect requires excellent prediction of the dimensional changes plus some means of correction of errors.

Use of butterflies may eliminate the need to spin the lens, thus avoiding the more serious stretch effects.

Construction can be done in Earth orbit, and the lens moved to its operational orbit when completed.

Lens Operations

The lens will not likely function as a simple, passive device, but will instead require constant attention to meet its orbital motion, attitude control, and focusing requirements.

The beam focal point should always lie ahead of the sail such that the beam is captured by the sail to the greatest practical extent, but with an intensity distribution as uniform as practical across the sail. This corresponds to keeping the beam diameter approximately equal to the sail diameter at the location of the sail.

The geometry of the laser, the lens, the ship, and the position and motion of the destination star impose pointing requirements on the system as a whole. These requirements cannot be met by tilting the lens because of the requirement to maintain the Δr spacing relative to the beam center.

If the lens is rotating, it is possible for it to experience some dynamic excursions. Flapping of the lens surface is a potential mode that could violate Δr spacing requirements.

For the example design of a 1000 km lens with the lens rings maintained in a fixed perfect geometry, the Δr spacing error will exceed the 0.56 m tolerance at the outer edge if the lens should tilt more than 0.09° relative to the beam. Beyond this angle, the beam will gradually lose its focus. Determination of an operationally acceptable limit angle on the lens will require an extensive analysis.

Lateral errors in the beam position from the ship's position are obviously of critical importance. The pointing objective is to aim the laser at the place the star will be when the ship arrives, and to keep the lens and ship on that line. Any motion of the laser emitter around the Sun seriously complicates the operation.

If the laser is in solar orbit or on Mercury, the lens will need to move in a halo orbit around the reference line shown in Figure 7.3. The diameter of the halo increases during the voyage of the ship, such that when the ship is well beyond the solar system, the diameter of the halo equals the diameter of the orbit of the laser. If the laser orbit is inclined to the reference line, which it would be if the laser is placed on Mercury, then the halo orbit must be elliptic rather than circular.

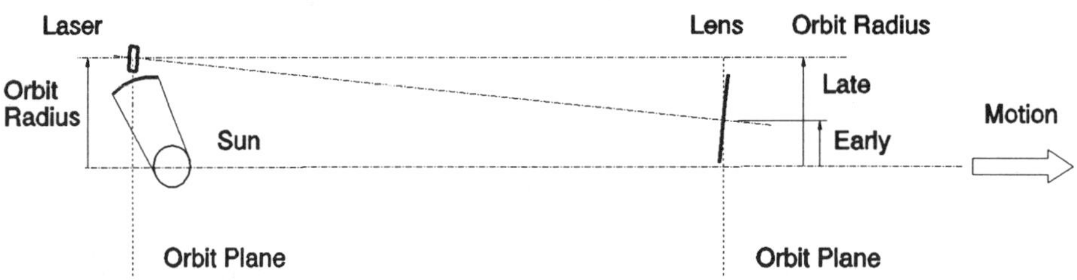

FIGURE 7.3 Orbits of the Laser and Lens.

The attitude requirements may not allow the lens to be tilted substantially to correct or control its position in space. If this is the case, some auxiliary attitude and position control system may be necessary.

The full consequences of transient failures to meet some or all of the position and pointing requirements need to be understood. It cannot yet be known how well future technology could meet these requirements, so the issue of the practicality of controlling the lens cannot be fully addressed.

Using a set of agile butterfly sailers as shown in Figure 7.4 may meet both

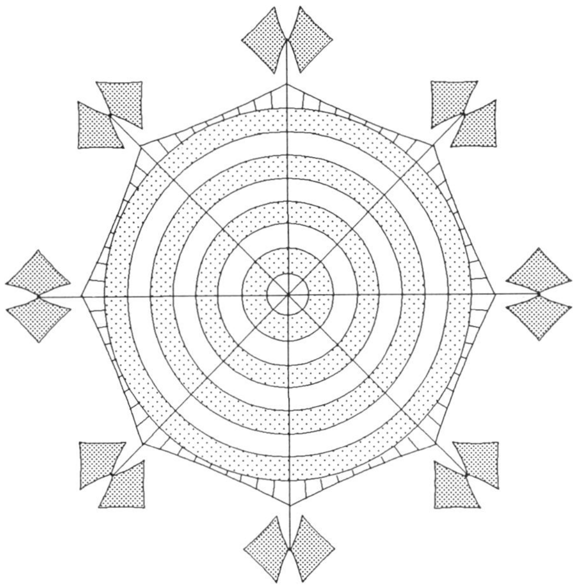

FIGURE 7.4 Butterfly Control of a Lens.

attitude and orbital motion control requirements. This technique can be used with the lens rotating or not. The sails will provide the tension necessary to maintain the lens shape if the lens is not rotating.

The beam lens is located further out in the solar system, between Saturn and Uranus, 15 AU from the Sun. It must be kept in the beam created by the conversion or reflector unit, which places significant requirements on the control of its motion. The lens must counteract the solar gravitational pull by using part of the beam energy for stationkeeping, or its motion must be adequately compensated for during the period of operation.

Allowing the lens to fall in toward the Sun could be acceptable provided the optical quality does not degrade to unsatisfactory levels during the operational period, which could be up to about 100 years. This inward motion could be moderated, if not fully counteracted, by using some of the beam energy on the lens.

The beam should be pointed accurately and stably during the period of operation. The system must correct for laser and lens motions and possibly for off-axis conditions.

The lens must be on the line from the source to the ship, corrected for light transit time (the line to where the ship will be when the light arrives). The lens should also be nearly perpendicular to this light line.

The beam should be pointed directly at the destination if it is the only beam in operation. However, if another beam is in operation at the destination, they both should be slightly offset from their opposite stars. This offset allows one beam to be used for acceleration and the other for deceleration, without counterproductive interference between the beams. The ships using the beams will make a lateral movement to go from their accelerating beam to their decelerating beam near the midpoint of the trip.

SHIP DESIGN

Interplanetary power beams can be used with solar sailing ships. An ordinary solar sailer can ride a laser beam, and could be used effectively in any region of a solar system.

The operation of an interstellar ship is essentially the same as for a solar sailing ship except for peculiarities in the configuration. The requirements for turning rates under normal conditions are much reduced.

An interstellar voyage with a substantial payload will require a very large sail and a low-mass design. Dimensions will be in the hundreds of kilometers for a manned interstellar ship.

The minimum payload needed to sustain the settlers at their destination is a factor that drives the design. This minimum mass is assumed to be 5000 metric tons.

The people also need to have an artificial gravity environment. A spinning disk sail is the logical choice, with a toroidal structure as shown in Figure 7.5.

The sail should fill as much of the disk area as practical to maximize the use of the power in the beam.

A lightsail designed for riding in a laser beam will probably be continuously exposed to intensities several times greater than the solar intensity at 1 AU. The sail temperature can be a limiting factor for the acceleration capability of the ship. Therefore, the absorptivity at the wavelength of incident light and the emissivity at the sail temperature are important parameters. Sail areal mass must be considered when coatings are used to enhance these parameters. If a leeside coating is used, the allowable

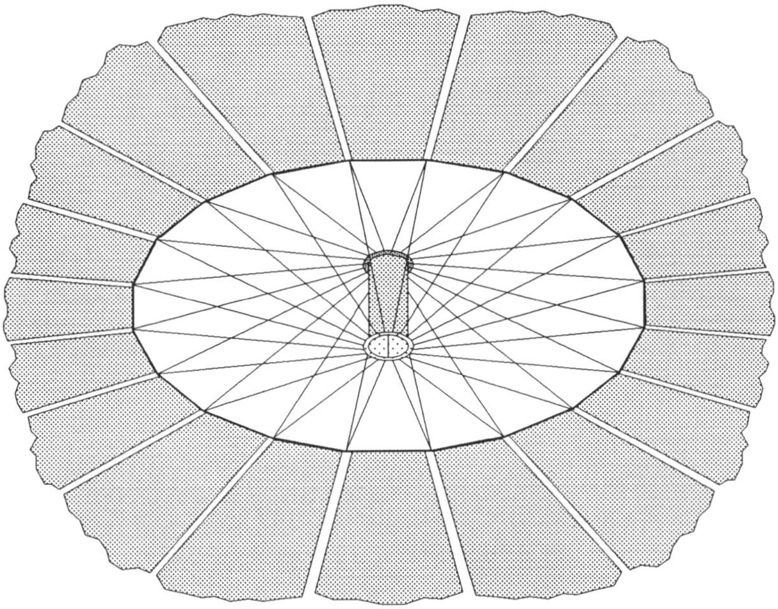

FIGURE 7.5 A Toroidal Hull for a Disk Sailer.

energy density can be increased, but the leeside emissions and the coating mass both work against ship performance.

A microwave sail is designed for longer wavelength electromagnetic radiation, and so can have a mesh design with mesh openings much larger than could be used in a lightsail. A lightsail can have holes in the sail to reduce mass if the holes are smaller than the wavelength of the light.

Spinning disk sails are probably tolerant of a substantial non-uniform pressure distribution across the sail provided that it is axially symmetric and the incident radiation is axial. A non-symmetric pressure distribution, if not corrected or compensated for immediately, will cause precession of the ship and sail, with a lateral translation resulting from the tilt of the sail.

In the case of a multi-stage ship, the back reflector can be either a parabolic or flat Fresnel zone plate reflector. The parabolic design relies on its shape to focus the back reflected light onto the second stage. The best way to obtain the necessary curvature in the reflector may be to have the surface made in sections held together by adjustable lines that control the billow in the surface.

The zone plate reflector, as with the lens, relies on phase shifting of the incident light to focus on the second stage. The zone plates would be one-half the thickness of the zones on the lens, because the zone plates here function as second surface mirrors, allowing the light to pass through the plates twice.

Both types of back reflector would continue to rely on spin with a weighted rim to keep the surface open. They both must be mechanically altered to change the focal point of the back beam. Neither type will be optically perfect, so they must be sized to compensate.

The acceleration of the ship can be determined from the expression

$$a = \frac{2P\eta}{Mc} \tag{6.4}$$

for power P, sail efficiency η, ship mass M, and light speed c. The use of constant acceleration and deceleration requires an acceleration capability of 16 mm/s^2 for a 100 year voyage to Alpha Centauri.

Some example concepts of interstellar lightships are shown in Table 7.1. Where there is more than one stage, each stage is included. The beam power levels are based on full use of the beam power, without safety factors applied.

The first two concepts shown in the table are robotic ships suitable for survey work. The profile of the voyage for the first and second concepts are shown in Figures 7.6 and 7.7, respectively.

The third example design, a ship having a 5000 metric ton hull and load combination, is sized for a one-way trip to Alpha Centauri. A two-stage sail is used. This ship cannot return to Earth, so the crew members reaching the destination become the beginning of a permanent settlement. This seems likely to be near the minimum manned voyage.

The fourth example is a three-stage ship intended for a round trip voyage, to be done within the lifespans of the crew. Its profile is shown in Figure 7.8.

SHIP OPERATIONS

The entrance of the ship into the beam must be done avoiding major intensity gradients across the sail. This could be achieved by the ship entering the beam very close to the lens, or having the beam unfocused. The ship could enter the beam with the sail turned at an angle to reduce the gradient effects.

TABLE 7.1 Interstellar Lightship Concepts.

Voyage	αCentauri[a]	αCentauri[a]	αCentauri	ϵEridani[a]
Type	Transit	Stop	Stop	Return
Stages	1	2	2	3
Payload, ton	0.33	26	5000	2900
Trip Time, yr	40	41	100	23+23
Mass, ton	1	71	7500	785
(by stage)		785	367,000	7850
				78,500
Diameter, km	3.6	30	129	100
(by stage)		100	400	320
				1000
σ, g/m^2	0.1	0.1	0.57	0.1
(by stage)		0.1	2.92	0.1
				0.1
Beam Power, TW	0.065	26	534	430
(by stage)		7.2	540	4300
				43,000
Lens Diameter, km	1000	1000	1000	1000
Lens Mass, ton	560,000	560,000	560,000	560,000
Maximum Speed, *c*	0.11	0.21	0.09	0.5
Average Speed, *c*	0.11	0.1	0.043	0.5

[a]Forward, 1984

This operation requires a carefully planned trajectory taking advantage of the large size of the lens relative to the sail. Once the ship has been positioned on the beam center, the beam can be focused onto the sail.

With the ship on the beam center and moving in a straight line toward the intercept point with the star, the operations will remain generally constant while the ship accelerates, except for refocusing the lens. If there is a coasting phase, the ship may be kept in a low intensity beam to provide operating power for the ship's systems.

If a two-stage operation is being done, the separation and reconfiguration will be a critical maneuver that must be done without any possibility of realtime coordination between the ship and the laser beam system. It will probably require great care to maintain the relative positions of the back reflector and the ship to avoid problems from shadowing and beam misalignment.

The deceleration phase using a back reflector will be even more complex

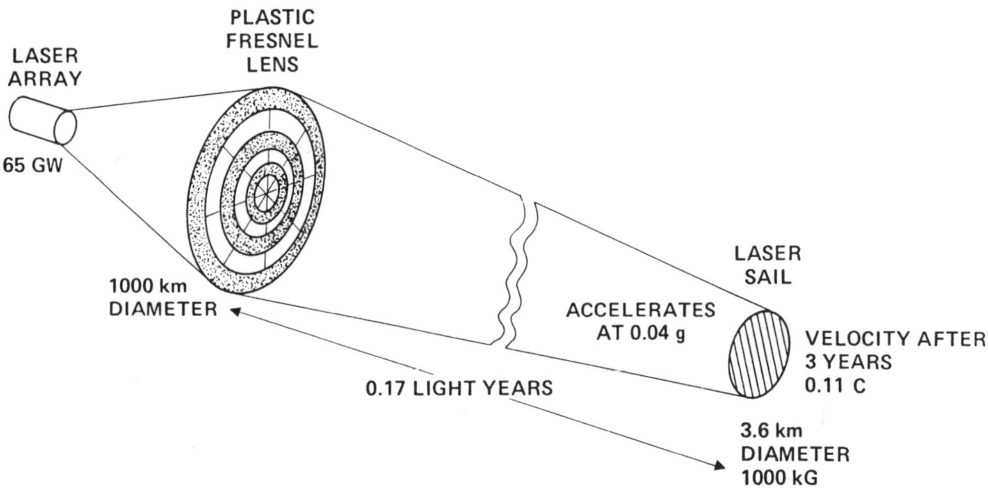

FIGURE 7.6 Profile of an Interstellar Fly-By Probe. (R.L. Forward)

than the acceleration phase since there are now four active elements in the system: the emitter, the lens, the back reflector, and the decelerating ship. The beam geometry during this phase is illustrated in Figure 7.9. The second reflection beam can be directed through the opening in the back reflector or dispersed so that little of the energy impinges on the back reflector. It can be seen that the primary beamed power works against the decelerating illumination by impinging on the lee side of the sail. Placing the ship to the side of the main beam would impose a lateral force of the back reflector, which would probably make this an unsuitable alternative.

If the upstream ship casts a shadow on the downstream sail, or reflected light impinges on the downstream sail, substantial deformations will result in the sails and serious torques will be applied to the ships. These interference conditions must be avoided, or an effective means of compensation must be devised.

The back reflector must deliver most of the beam power to the subsequent stages, allowing for a safety factor. The difference in sizes between stages requires that the reflected beam be focused down to the next stage.

With the ships operating colinearly as shown in the figure, the cutout in the downstream sail will need to be larger than the upstream sail to provide a lateral safety factor.

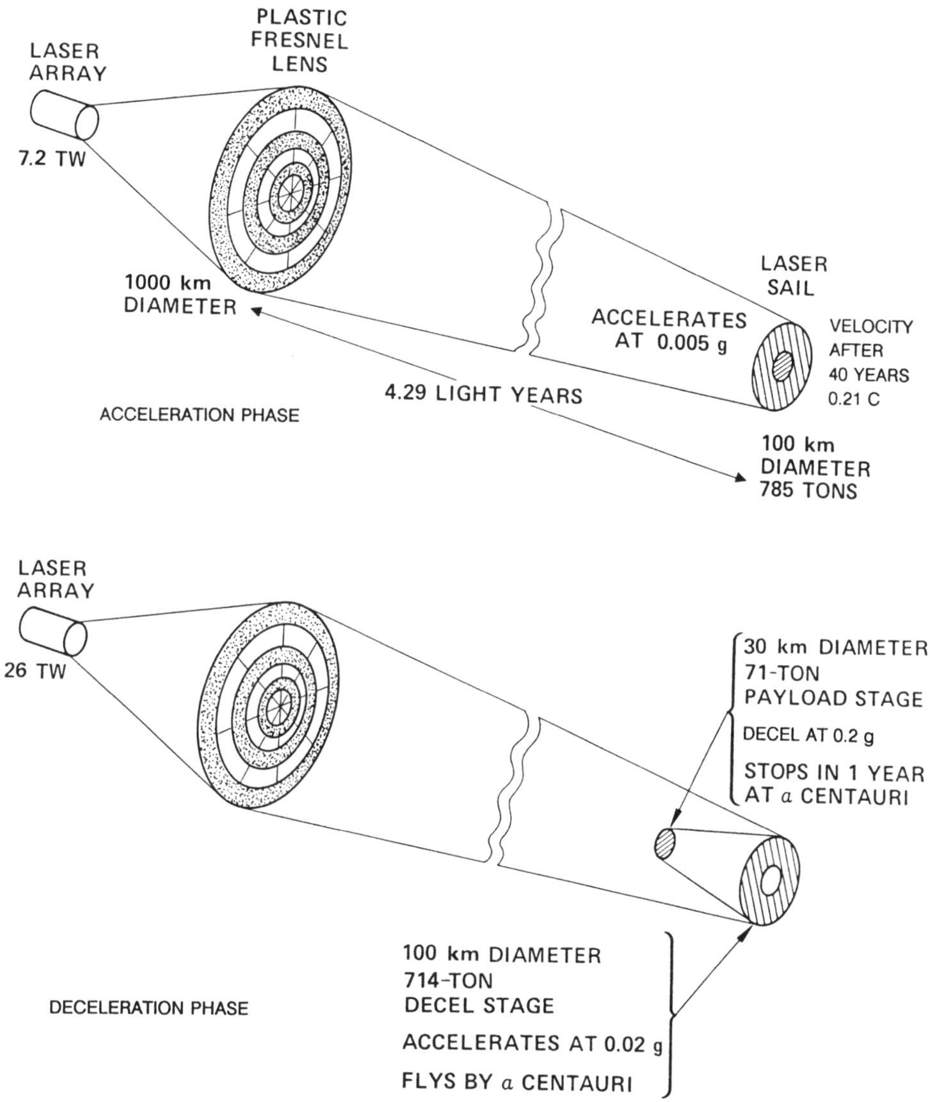

FIGURE 7.7 Profile of a Voyage to Alpha Centauri. (R.L. Forward)

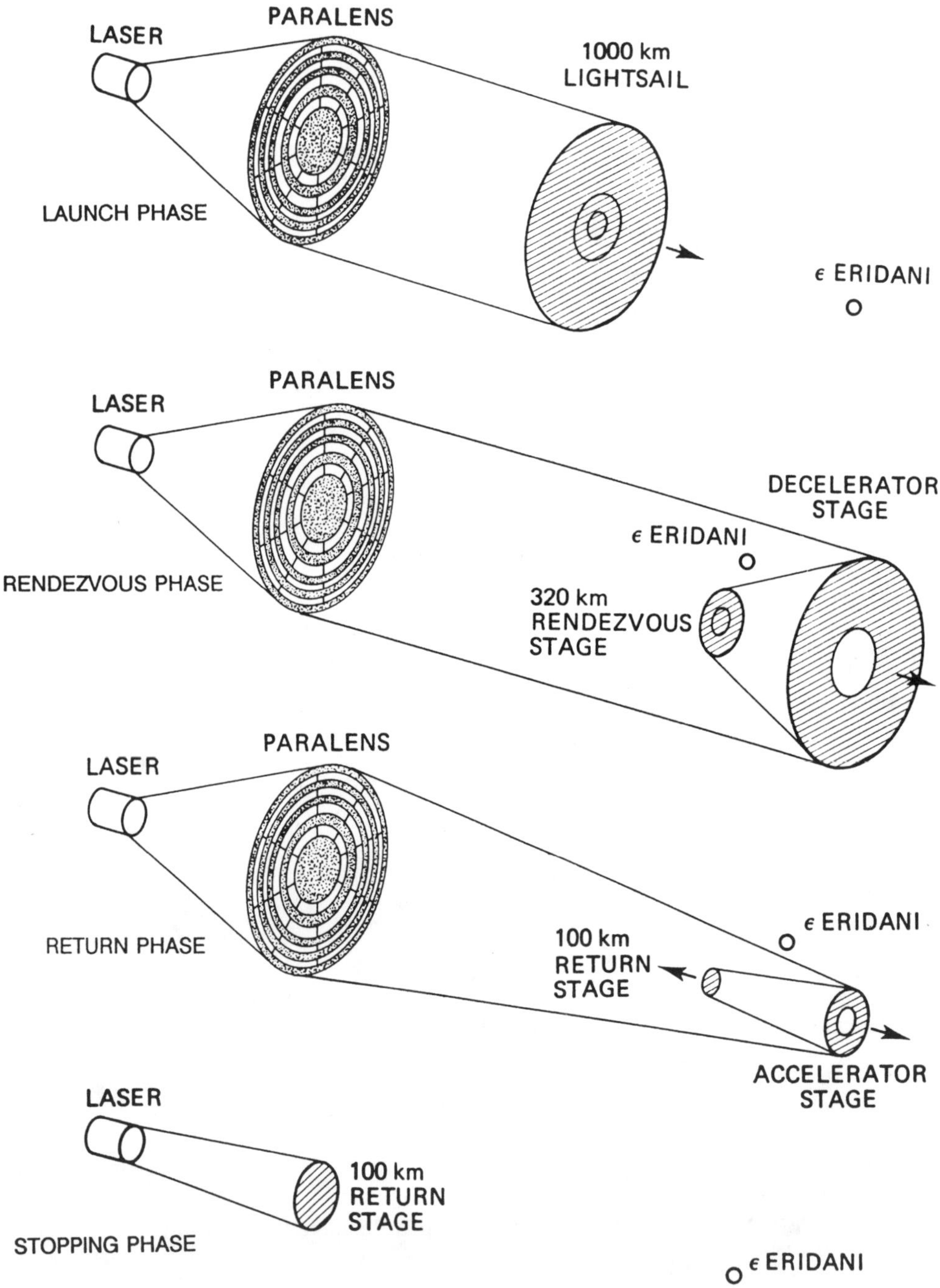

FIGURE 7.8 Profile of a Roundtrip Voyage to Epsilon Eridani. (R.L. Forward)

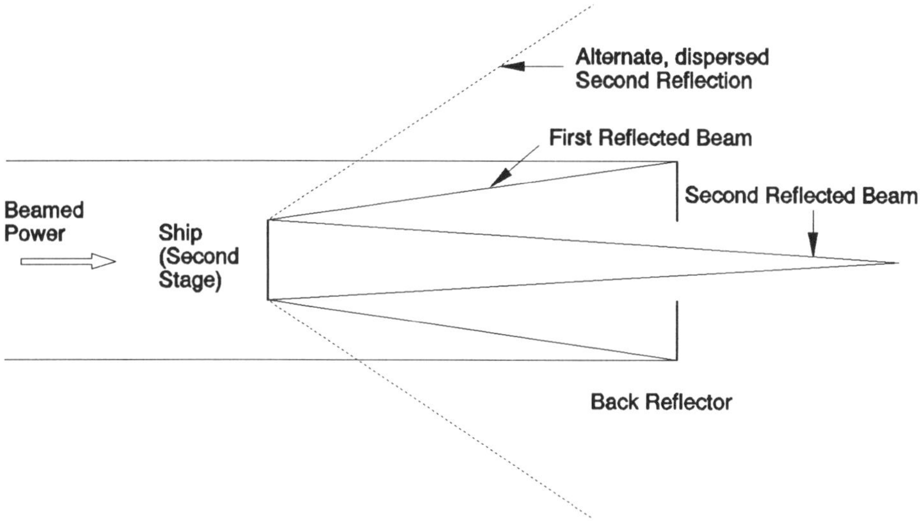

FIGURE 7.9 Geometry of Power Beams and Sails During Deceleration.

The quality of the reflected beam will be less than the original beam due to imperfections in the reflector. The energy impinging on the backside of the decelerating ship increases its sail temperature and reduces its deceleration capability.

The arrival maneuver must allow for an error between the beam position and the star position. The ship must be driven out of the beam in a direction toward the star to enter a desirable orbit. The ship should retain its sail, but the beam departure must be done without losing control of the ship's attitude or substantially deviating from the desired direction.

An alternate strategy is to send a decelerator ahead of the manned system using a two stage system. The sail that arrives at the destination would serve as the back reflector for subsequent ships, which could be single stage ships. If this is done, it would make sense to include a substantial set of robots able to provide maintenance, survey the stellar system, and even collect resources before the arrival of the settlers.

Multiple beams can be simultaneously operated for different destinations. The beam units reside in orbits peculiar to each beam if they are placed in solar or mercurian orbits.

Ship Guidance and Control

Navigating and guiding a ship along a beam extending from one star to another is not a trivial task in spite of the apparently simple geometry. Some key factors of the problem are:

- the telecommunications time lag prevents coordination between the ship and the beam
- the tight beam focus requires great precession in guiding the ship
- failures in the sail structure could create serious unbalanced torques on the ship
- the beam must be pointed at the computed position of the destination star decades in the future, based on the estimated arrival time of the ship.

The speed of light places a constraint on communications between the ship and the laser station creating the beam. If an interstellar ship drifts away from the center of the beam, the beam cannot track the ship to assure full illumination while a course correction is made. The burden is then on the ship to guide itself along the beam, remaining close to the centerline.

The ship could maintain a light beam carrying navigation and other information back to the laser station. This could be useful in making some adjustments in the pointing of the beam, but with a limitation because of the delay in the receipt of the data and propagation of the modified beam.

Intensity sensors deployed at or beyond the outer edge of the sail are probably essential to aid in measuring for the location of the beam center and to warn of a close approach to the beam edge.

If a ship drifts beyond the edge of the beam center, there may be hope for recovering the ship using the light that spreads outside of the beam center. This will depend on the ship having only a small lateral velocity and being brought quickly under control.

Lateral motion of the ship can be controlled by tilting the ship at a small angle to create a lateral force in the desired direction. This must be done carefully to avoid losing control of the ship, which could result in leaving the beam.

Another approach is to use lateral thrusters. They would provide a rapid, easily controlled lateral force, at the expense of having to carry the extra mass of the system. These approaches can be used together.

The avoidance of large bodies is an important consideration. With a beam width near the diameter of the sail, little could be done except to make a lateral movement to protect the living quarters. The sail would be damaged in such an encounter, which could require compensation by the control system if a significant pressure imbalance resulted.

These problems of staying on the beam and pointing correctly are worse if a reflected beam is used. Then two remote objects are to be considered, along with a much greater distance.

The first ship to a destination has the most difficult navigation and guidance task since there will be some uncertainties in the beam characteristics, the distance to and position of the star at the time of arrival, and some uncertainty in the actual average speed of the ship. The navigation uncertainties add to the value of making the first interstellar beam rider a robotic ship.

MICROWAVE SYSTEMS

This system is misnamed since its operating wavelength is in the tens of millimeters, while the laser system actually operates at or near one micrometer. The radiation of the beam is fundamentally the same as that of light, differing only in the wavelength. However, this difference does result in some major design differences in the beam system and the ship. The data presented here on the Starwisp concept are from Robert Forward.[3]

The Microwave Source

Solar power satellites may become operational in the next few decades, with proposed output levels of 10 GW of microwave power in the X band ($\lambda = 30$ mm). Their radiating antennae would have diameters near 1 km. These satellites could be attractive sources of energy to drive small interstellar probes if the economics of the project are reasonable.

The satellites would be located in geostationary orbit around Earth. Their microwave antennae are steerable, so that one could be readily directed to a remote lens that would focus the energy on a small probe.

The Microwave Lens

The best type of lens for focusing the microwave beam is one made of wire mesh in alternating rings of mesh and open space, creating a Fresnel lens with phase reversal zone plates. It functions like the zone plate lens for a laser system, but is tuned for the microwave wavelength.

An example design was defined with a resulting diameter ratio of D/d equal to 50,000, with a sail diameter d of 1 km. This required diameter ratio decreases substantially if the final velocity v is decreased from the design value of 0.2 c.

Starwisp Operations

The Starwisp driver is a 10 GW microwave beam, initially accelerating the spacecraft at 115 gravities. The craft reaches 0.2 c in only a few days, and can then coast the remainder of its 21 years to Alpha Centauri.

The same beam could be used to send other Starwisp probes to other stellar systems or regions of interest, so that a fleet of these craft could be running through interstellar space. These craft would be valuable in helping us understand the nature of stellar systems and other structures and phenomena in our galaxy.

Long range probes could be sent to study gas and dust clouds, formative and terminal stellar events, and even the structures and processes in the center of the galaxy.

When each probe reaches its destination, it would again be illuminated by the power beam so that it can make its observations and transmit its information back to the solar system.

This approach allows a single power beam to drive numerous probes that could potentially have very low costs considering the value of their accomplishments.

A solar power satellite in geostationary Earth orbit could be used to drive a Starwisp probe, Figure 7.10. The capital cost of the satellite would not be part of the probe project. Instead, only a few days of output would be needed to send each probe to its destination. The cost of the electrical energy used to drive the Starwisp to its coasting speed is about \$50 to 100 million, based on the operation taking a power satellite's output for only a few days.

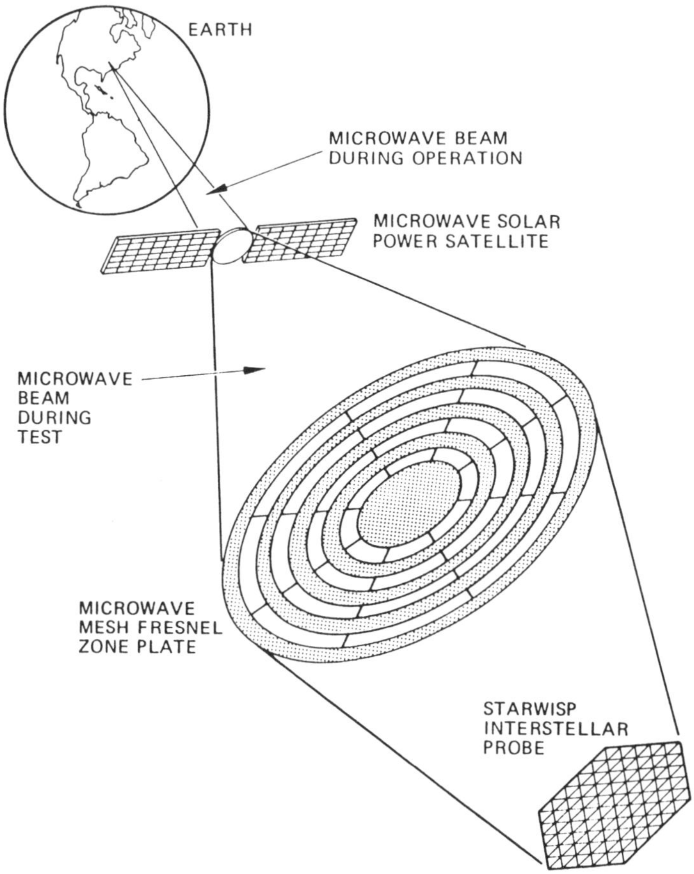

FIGURE 7.10 Operation of a Starwisp Probe. (R.L. Forward)

Starwisp Guidance and Control

Trajectory corrections are essential to assure an accurate arrival at the destination. The probe must be able to change its attitude and observe its pointing direction to enable these trajectory changes.

Periodically along the trajectory, course corrections will be made based on tracking data obtained by signals reflected off of or transponded back from the probe. When the trajectory correction impulses have been determined, the probe must be oriented for the maneuver. The orientation should be confirmed by having the probe view the target and other stars, and then send the information to Earth. Once the orientation is confirmed, a power beam will deliver a precise amount of energy to the probe. Several course corrections will be needed over the trajectory. The difficulty of making the corrections will increase during the voyage as the round-trip communications time approaches 8.6 years.

The trajectory changes might be made without waiting for attitude observations from the probe if the probe has a memory, a knowledge of star patterns, and an ability to make autonomous attitude changes. If the probe is considered to be sufficiently reliable for this, a corrective maneuver could be done by sending pointing instructions to the probe, followed a short time later by the corrective impulse. Subsequent tracking data would provide information on how well the correction was done.

REFERENCES AND NOTES

1. Forward, R.L. (1984) Roundtrip Interstellar Travel Using Laser-Pushed Lightsails. *Journal of Spacecraft*, Vol. 21, No. 2, March-April.
2. High Powered Lasers in Space. *Astronautics and Aeronautics*, Vol. 17, March, 1979.
3. Forward, R.L. (1985) Starwisp: An Ultra-Light Interstellar Probe. *Journal of Spacecraft*, Vol. 22, No. 3, May-June.

8. SCENARIOS OF THE FUTURE

Five scenarios are discussed in this chapter, selected because they may be the most pivotal in the development of human civilization in space. Sailing ships are described as the major form of transportation in each scenario, but other approaches can certainly be used. However, the sailing ships provide a benchmark of performance considering factors such as payload capability, transfer time, development costs, operating costs, and reliability. If something else can be found to outperform sailing ships in these important factors, then we are even better off than before.

The values presented on transfer times, payloads, and sail size are arbitrary to an extent. They show things that can be done, within the accuracy of various assumptions, not the way things ought to be done. The engineers who plan the voyages and design the ships will find the best solutions within whatever constraints they must work with.

SAILING NEAR HOME

The first clipper ship will likely begin as components built on Earth and carried into orbit. There the parts will be put together at a construction platform located on or near an operating orbital station. The structural members and the sail will be deployed. Following a thorough checkout of the operating systems, the ship will be elevated on a roughly 300 km tether and released into an elliptic orbit to begin sailing. Its shakedown cruise might include a swing around the Moon.

The ship may not be meant to travel near the Sun, so the stress levels in the sail and supporting structure will be low. The construction platform can provide temporary devices to deploy the sail, avoiding having the ship's structure carry the deployment loads. The supporting structure will then be very light, so with a thin, aluminized plastic sheet for the sail, the ship will have good sailing performance with a sail loading of approximately 3 to 5 g/m^2 without payload.

When sailing ships begin routine operations in Earth-Moon space, illustrated in Figure 8.1, they might likely be operating on a commercial basis, with shipping companies letting the ships out on contract for customers purchasing transportation.

An obvious candidate for application is the lunar development effort that is intended to start at the beginning of the new millennium. To some extent this scenario is redundant with the one for the development of Mars. The overall pattern of transportation is similar except that transfer times between Earth orbit and lunar orbit are shorter than to Mars, and transfer opportunities are more frequent. The use of sailing ships in lunar development would require people to rethink the time scales involved because the transportation time would be quite long compared to ballistic transfers.

Sailing ships can readily carry supplies, equipment, and facilities, allowing a substantial savings on transportation costs. Outbound cargo is picked up at a dock in low orbit, the ship is elevated by a tether, and then spirals across to the Moon where it descends into a low orbit. The cargo is delivered to a lunar orbital station, where it is placed on landing craft that takes it to the surface, or the payload package may take itself to the surface. The ship will take on any Earth-bound cargo and begin its return trip to low Earth orbit.

A 1000 m ship could transfer a 4.4 ton load from low Earth orbit to lunar orbit in about 180 days. There, the payload's own propulsion system could put about 1.7 tons on the lunar surface, or a lunar ferry could carry about 3 tons to the surface using lunar oxygen as one of its propellants. Either way, the cost of getting the package from the surface of Earth to the lunar surface might not be more than $50 million using a commercial launcher on Earth. Getting a 1.7 ton package from Earth's surface to lunar surface using the Space Shuttle and chemical propulsion systems would cost the taxpayer about $400 million.

Sailing ships can also be used to support stations and development projects at the various Lagrange points, where stable orbits exist at remote distances from Earth. The ships can provide transportation between these sites and

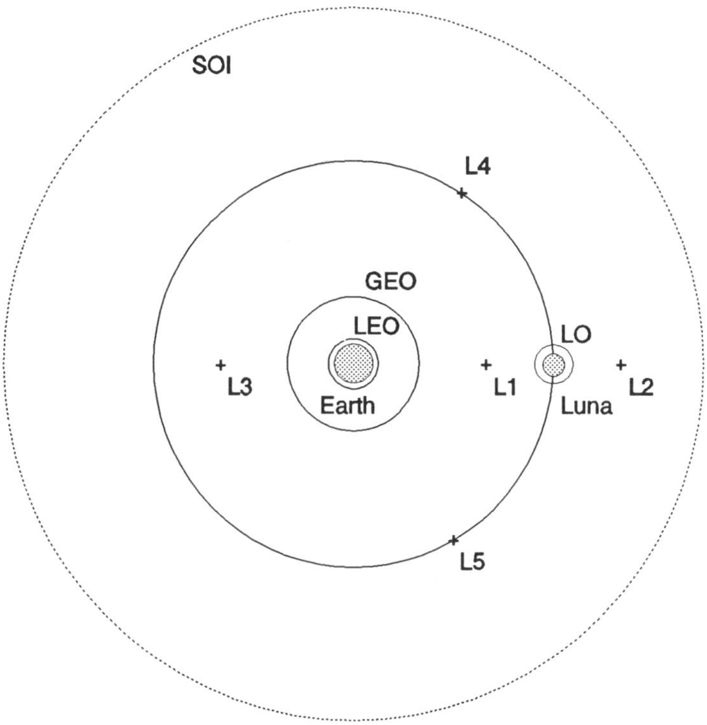

FIGURE 8.1 Sailing in Earth-Moon Space. Low Earth orbit, geostationary Earth orbit, lunar orbit, the Lagrange points, and the sphere of influence are shown as locations accessible to sailing ships.

Earth and lunar orbits at costs far below those for chemical propulsion.

Ferrying between low orbit and geostationary orbit might become a major operation for sailing ships. At present, the transportation cost for geostationary communications satellites is about equal to their production costs. Sailing technology could possibly reduce the unit transportation cost by a factor of four to five. A 500 m ship could carry a 1.1 ton load into geostationary orbit in about 90 days.

Sailing ships could be ready for operations by the year 2000 if a market develops for them. They could be operating in time to support the return of people to the Moon.

CITIES ON MARS

Sailing ships could make travel to Mars an economical, practical operation. This is critical to establishing not just a human presence on Mars, but more importantly, a successful, permanent social system with a complete infrastructure of cities and all the other necessities of civilization.

The leading contenders for transportation to Mars in the current NASA studies are irrelevant for commercial and personal transportation. Chemical and nuclear propulsion are so expensive that only the government can afford them, at least in the coming decades.

A 2000 m sailing ship can carry a manned module with a mass of about 20 tons to Mars in 4 to 5 months, depending on the year of the voyage. This is the size module being considered for the first exploratory expedition.

Operations with sailing ships eliminate massive quantities of propellant that are required by other means of transportation. Each ton of propellant eliminated from an expedition is a savings of $5 to 10 million. Mars missions by chemical propulsion use hundreds of tons of propellant, leading to billions of dollars being spent just to carry the propellant from Earth into orbit. Additional savings will result from the short transfer time for the crews, which reduces consumables, could allow smaller quarters, and removes the concern over lack of gravity during a long transfer. The latter eliminates the possible need for transfer ships providing artificial gravity.

The full reusability of a sailing ship allows amortizing its initial cost over several voyages, further reducing the cost of any single mission. No other type of space transportation can fully offer this advantage.

This low-cost transportation system may allow a more rapid buildup of population and infrastructure than any other approach. As a result, we could see cities existing on Mars in the second half of the Twenty-First Century.

Survey and Sample Return

The Mariner and Viking spacecraft provided an important beginning, but the survey of Mars is far from complete. Mars orbiting spacecraft are needed in substantial numbers to support a human presence on the planet. Once a sailing ship is constructed, its main role could be transporting survey devices to Mars orbit. With the high cost of lifting loads into Earth orbit, one or two trips to Mars by the first ship might pay for the full development cost of the ship out of the savings from not carrying propellant into Earth orbit.

The ship would leave from a moderately low Earth orbit with a survey device. After completing the escape spiral, it would sail to Mars along one of two possible types of trajectories. Depending on the relative cost impacts on the survey device and the value of returning the ship promptly for another mission, the survey device would be carried into a Martian orbit or released near Mars and allowed to execute its own capture maneuver. These options exist for both orbital and surface survey devices.

If the ship goes into Martian orbit, it will be gone from Earth for roughly 4 years, counting two spirals at Earth and two at Mars. If the ship only swings around Mars and makes an immediate return to Earth, it would arrive back in low Earth orbit approximately 2 years after its departure. If the round-trip time does not exceed about 2 years, the ship could pick up another load in time for the next good departure opportunity. This process can go on repeatedly, with the same sailing ship carrying a load to Mars at each good departure opportunity.

One of the survey devices could be a surface rover that operates for many months on Mars, Figure 8.2. In addition to making observations and analyses, the rover picks up diverse samples from a variety of sites during its travels. After locating a good landing site, the rover provides a guidance beacon for a lander that receives the samples for transporting to Earth.

The early return of samples from Mars could greatly aid in planning for the eventual presence of humans on Mars. A round trip to the surface of Mars using chemical propulsion requires a large quantity of propellants relative to the craft landed on Mars and the sample brought to Earth. This is true even if some of the propellants are obtained by manufacturing processes on the surface of Mars. A sailing ship can provide all of the interplanetary transportation for the mission, with chemical propulsion used only for the Mars landing and launch back into orbit. This can drastically reduce the mass initially needed in Earth orbit, and provide a major savings in the cost of the project.

A square sail 700 m on a side could carry a 4 to 5 ton lander and ascent vehicle and return with an estimated 150 to 200 kg of samples, shown in Figure 8.2. This compares against 20 to 40 kg of samples from a larger, more costly chemical propulsion approach. The total mission time would be about 3.5 to 4 years including the departure and arrival spirals at Earth.

FIGURE 8.2 Mars Sample Collection. (JPL)

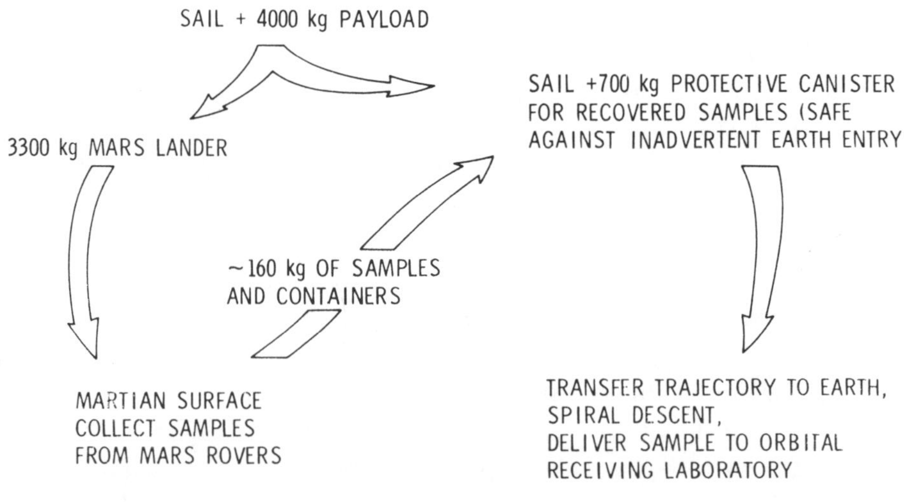

FIGURE 8.3 Sample Collection Profile.

The First Manned Mars Mission

Sailing ships could have a major role in the first manned expeditions to Mars. The ships can influence the efficacy, safety, and cost of the expeditions. One possible strategy is to use sailing ships to carry cargo, and then to consider sailing, chemical, or other propulsion modes for transporting people.

Long before the departure of the crew, sailing ships can be making routine runs to Mars carrying such things as pressurized station modules, propellants, life support supplies, landing craft, and emergency return spacecraft. A major cache of materials and supplies can be built up in Mars orbit and even on the surface before the crew actually departs from Earth.[1] The resulting base on the surface can be very well equipped, as illustrated in Figure 8.4.

A 1000 m sailing ship could carry 5 to 10 ton loads, and a 2000 m ship could carry 20 to 40 ton loads. Their round trip times using swing-by encounters at Mars is approximately 2 years including two Earth spirals.

A small fleet of sailing ships uses each Earth departure opportunity to make the equipment buildup. The first loads are items that can be delivered early in the program, such as propellants and life support supplies. Next is the sophisticated hardware such as landers and habitable facilities.

The early delivery equipment includes survey devices used for the final

FIGURE 8.4 Martian Base Supported by Sailing Ships. (JPL)

landing site selection. This can include high resolution orbital cameras, camera-equipped aircraft, and surface rovers. The rovers will make inspections of the final candidate sites and will then remain to serve as guidance beacons for the landers.

Precursor landers at the selected site can include life support supplies, instruments, tools, reserve propellants, and habitable volume. A supplementary ascent vehicle can be placed on the surface for emergency use by the crew. Then if the primary ascent vehicle launches successfully with the crew, the supplementary vehicle can be used for carrying additional samples to orbit.

Equipment placed in orbit constitutes the beginning of a permanent orbital station. This precursor equipment includes life support supplies, reserve propellants, and habitable volume. The manned landers can also be placed in orbit with the station to reduce the mass carried by the manned interplanetary vehicles.

This approach results in the delivery of large quantities of materials which exist in known locations or orbits, and whose operating conditions are known through telemetered data. This verification of delivery and condition greatly reduces the risk for the crew, especially considering the potential for having more emergency equipment than could possibly be carried on the same vehicles as the crew.

There is also an additional major benefit to the crew and the efficacy of the mission. Since so much verified equipment is already in place, the crew vehicles can travel with a minimum of equipment for the general expedition, concentrating instead on life support supplies to provide for crew vehicle contingencies. Since the crew vehicles are carrying minimal mass, their transfer times to Mars can be greatly reduced. This reduces boredom and stress fatigue associated with the otherwise long transfer to Mars, and could essentially remove the concern over extended time under zero gravity conditions. These benefits can be realized whether the crew transfer is by sailing ship or chemical propulsion.

If chemical propulsion is used for crew transfer, these benefits can be further extended by using the propellant cache already in Mars orbit. This propellant can be used to reduce the return transfer time, although the crew vehicles should have enough onboard reserves to make the long return to Earth as a contingency. If at the time of the expedition enough confidence exists in the sailing ships, the ships can be used for making fast transfers with the crew vehicles. The outbound trips are shorter than the fast transfers by chemical propulsion. The initial mass in Earth orbit would also be greatly reduced.

At the beginning of the expedition, the crew vehicles would be boosted to high Earth orbit where they would be attached to the sailing ships, avoiding the Earth spiral time for the crew. The sailing ships would take as little as 4 to 5 months to reach Mars. The crew vehicles would separate from the sails on arrival and use aerodynamic braking to enter Mars orbit in the same manner as used for cargo delivery. The precursor cargo deliveries could have been used to provide verification of the crew vehicle designs and operating techniques.

The crew vehicles can abort the mission before braking and make a direct return to Earth even after separating from the sailing ship. Each crew vehicle has sufficient supplies to make the return to Earth on its own. If an abort did take place in this manner, a sailing ship could be sent to the crew vehicle to assist the return. This ship might be one from the fleet which carried the crews to Mars.

After a normal braking into orbit, the crew vehicles would maneuver to the station to begin preparation for the descent to the surface. These activities would rely on the materials brought by sailing ships using optimal transfer opportunities in earlier years.

At the conclusion of the surface exploration, the surface crewmembers would return to the station. The crew could then return to Earth using chemical rockets which had been placed at or near the station.

The return could also be made using sailing ships waiting in Mars orbit, possibly put into place before the crew arrived at Mars. For maximum safety, the crew could have both sailing and chemical means of returning to Earth. The ships would provide a flexibility which the chemical rockets would not be able to do in that they could leave Mars at any time to head back to Earth. The chemical rockets allow departure only in a departure window, when the propellant available is adequate for the energy requirements. If the return window is missed, the wait for the next one is two years if only chemical propulsion is used.

The sailing ships would provide a reasonable return to Earth, typically taking about 10 to 12 months. For return by either chemical or sailing methods, the crew vehicles would use aerodynamic braking to enter Earth orbit at the end of the expedition.

There are four major benefits from the use of sailing ships in this scenario: 1) the reduced time spent in transit between the planets, 2) the flexibility to depart Mars at any time after arrival, 3) the added safety from having the additional equipment, and 4) the cost savings resulting from not carrying large quantities of chemical propellants into Earth orbit and to Mars. The savings resulting from reduced launch operations probably exceeds $10 billion, even after allowing for the use of less costly launch vehicles than in current use. This savings exceeds by more than an order of magnitude the cost of developing and constructing a fleet of commercial sailing ships which could support the expedition.

Commercial Transportation to Mars

Exploratory expeditions will give way to commerce as permanent sites are established on the Martian surface and in orbit. The nature of transportation to Mars will change substantially when this happens. This process could be aided by using facilities and equipment remaining from the exploratory expeditions.

When people are routinely being carried to Mars, the ships will not have to include the large and massive quarters necessary for their stay at Mars, nor the shuttles needed between the surface and orbit. Shipping will rely on the existence of adequate facilities at Mars to support the people on arrival. This will allow reduced loads on the passenger ships, with the resulting transfers to Mars approaching 3 to 4 months.

The use of sailing ships to transport people and goods to Mars, or any other destination away from Earth, will hold down the cost of transportation and thereby aid in the development of space commerce. The cost of delivery from Earth orbit to Mars orbit by sailing ships will be a fraction of the initial cost to orbit. Using chemical propulsion, the Mars delivery cost increment is about five times the initial cost to orbit.

Mars appears to be able to provide abundant raw materials necessary to support the development of cities and a major social system. Water is abundant in the form of ice which will meet the direct needs of people and also provide propellant for vehicles operating between the surface and orbit. Since this propellant does not have to be carried from Earth, it will be relatively inexpensive, allowing the operation of low cost shuttles from the surface to orbit and back.

Commerce between Earth and Mars could be done using low-cost launch vehicles at Earth, low-cost reusable ascent/descent vehicles at Mars, and sailing ships crossing interplanetary space at the lowest cost of any space transportation system. This use of sailing ships appears to be the lowest cost scenario for moving people and goods between Earth and Mars.

DEVELOPMENT OF A COMET OR ASTEROID

Comets and asteroids can be important sources of materials for the construction of facilities in space. This is primarily because of the low energy requirements for moving material from these bodies to Earth. The small bodies can provide a wide range of materials, including silicates, iron, oxygen, water, carbon, nitrogen, and hydrogen.

These small bodies are also of scientific interest in addition to being reservoirs of raw materials. Many of the bodies have remained in nearly the same condition as when they formed shortly after the beginning of the solar system, potentially providing a wealth of information on the formative processes at that time.

A trip to the surface of another body in the solar system must include the energy requirements for descending to and ascending from the surface along with the energy requirements for crossing space to reach an orbit around the body. This makes some of the Earth approaching asteroids the easiest extraterrestrial surfaces to reach in terms of energy requirements.

A sailing ship can reach any body without propellant consumption, but the descent/ascent vehicle must use propellant. A round trip between Earth orbit and the surface of the Moon or Mars requires a substantial quantity of propellant even if sailing ships are included in the overall transportation.

A landing and return mission to a small body by means of a sailing ship can be done with almost no propellant consumption. This is important in keeping mission costs low.

Survey

The first visits to small bodies will be for making observations and returning samples for detailed studies. Observations will be made from instruments carried on the ship or a detachable spacecraft that can move in close to the surface. These observations can make precise measurements of size, shape, and mass plus more detailed observations of chemical elements, mineralogy, and geology.

A separable spacecraft could also descend to the surface for direct observations of surface and subsurface material. It might place sensors in several locations to measure heat flow and map the interior using shock waves. A spacecraft with these capabilities could also carry samples to the sailing ship for return to Earth.

The sailing ship, detachable spacecraft, and samples can return to Earth directly or after visiting one or more additional small bodies during the mission. Upon return, the ship would enter Earth orbit where the samples would be removed and the propellant supply of the lander replenished. The ship and lander could then leave on another similar mission. The key to success of this low-cost, multi-mission concept is that all ship and spacecraft hardware is reused and only minor propellant quantities are consumed. No other interplanetary spacecraft concept appears to have as much potential for low-cost exploration of small bodies.

Base Development

Advanced exploration and the establishment of mining operations can also make use of sailing ships. Similar to the situation for the development of Mars, sailing ships can, as a minimum, provide low-cost transportation of heavy loads. The ships could also carry people, robots, special supplies, and refined products.

The low energy requirements for transportation make the asteroids and comets approaching close to Earth potential sources of large quantities of raw materials for use in Earth orbit. Offsetting the relatively low transportation costs are the relative inaccessibility of the bodies except for periodic opportunities and the large capital costs of establishing mining operations on the bodies.

The bodies themselves might be turned into habitats, which could simply be an adjunct of the mining operations. As mining progresses through a body, the opened parts could be turned into habitation and farming areas supporting a large population of people and robots building large-scale structures and sending resources to other locations.

The bodies could serve as the source of materials for the construction of planetoids adjacent to the bodies. They can provide enough material for the construction of thousands of planetoids, having living areas ranging from that of a small city to a small country.

There are several reasons then for the possible establishment of a permanent base on one or more of the bodies. Sailing ships can provide a low cost means of enabling these developments.

DEVELOPMENT OF MERCURY

Mercury has the potential for being an important site for mining and materials processing operations to support large space projects such as interstellar energy beams and the construction of planetoids. It has a high concentration of metals and other non-volatile materials. Being close to the Sun, it also has abundant energy available to support the mining and processing activities.

Survey and Preparation

The energy requirements for reaching Mercury are so large that a major surveying effort is unlikely without some form of advanced propulsion which is also low cost. The solar sail seems to be the most suitable approach by far.

A ship with a 700 m sail could carry 5 tons into Mercury orbit, taking about 15 months for the interplanetary transfer. This is adequate to include an orbital survey satellite and a lander which could be placed at one of the poles. If Mercury is to eventually be used as an interstellar power station, the south pole would be the landing site because of the visibility of Alpha Centauri from there.

The landing equipment would include a rover, drilling and digging equipment, and instruments for mineralogical and elemental analyses. Samples could be launched into orbit for pickup by the ship and return to Earth. Including a sample return would reduce the amount of equipment on the surface or would add to the transfer time because the mass of the ascent rocket adds to the load carried by the ship.

Subsequent trips, which could be done with the same ship, would carry additional robotic equipment for a thorough exploration of the site. The site would also be prepared for a permanent base by the robots, acting on instructions from Earth. Eventually, habitable structures would be prepared, and a cache built up including life support supplies and rocket propellants. An ascent rocket suitable for people could be placed at the site. A permanent orbital station could also be established if it were of benefit to support manned operations.

Manned Expeditions

Unlike the case with Mars, chemical rockets are not a viable option for transporting people to Mercury. It would be difficult to find a method of transportation more effective that sailing for carrying cargo and people between Mercury and Earth in the next few decades.

The sailing part of the expedition would begin in a high Earth orbit, probably at the L4 or L5 points. Most of the necessary equipment would be en route to Mercury or already in place by the time the crew departs Earth. The manned ships would be lightly loaded, carrying only the crew and life support supplies sufficient for the round trip. This provides the shortest interplanetary transfer times, which will be roughly one year each way.

The ships will brake by solar pressure to enter orbit at Mercury. After maneuvering into suitable orbits, the sails will be turned away from the Sun so the ships will remain in their orbits. The crews will use chemical propulsion for descent to the surface site which has been established by the robotic work force.

The descent rockets are also capable of carrying the crew back into orbit, or they can optionally use an ascent rocket already in place. After perhaps 6 months at the site, the crew would ascend to orbit and leave the rocket at the orbital station. They would transfer to the sailing ship for return to Earth. The total mission time is then about 2.5 years, which is shorter than some of the manned Mars missions currently being considered.

Base Development

While the first crew is working on the surface, the cargo ships can return to Earth to pick up additional loads and possibly the next crew. The second crew could arrive at the surface site around the time of departure of the first crew. This same rotation could be done with subsequent crews.

Depending on the desired rate of buildup of the base, it could be possible to keep a more or less continuous manned presence at the site from the time of the first crew. The robotic operations would be continuous with or without people present.

The first expedition could then also be the beginning of a permanent presence on Mercury. A small permanent society could be established, initially sustained possibly by development of a power station for the interplanetary ships. Shipping would be a continuous operation to support the community and the power system.

Ships going to Mercury can carry very massive loads. The 1000 m ships can carry 25 ton loads on 18 month voyages, while 2000 m ships can carry 100 ton loads.

Long spirals out from an Earth orbital station can be avoided if several ships carry parts of the total load out to a remote point, such as L4 or L5, where a single ship is given the entire load. This ship could be one of the ships which spiraled out from low orbit or it could be a specialized interplanetary ship.

If the payloads are needed at Mercury in minimum time, the ships would run lightly loaded. This requires using more ships for the interplanetary transfer, but they all return in less time than a single ship carrying a large load between the planets. A single ship could also carry a large load all the way from low Earth orbit, but the total trip time might reach 6 years or more.

When the Mercury base becomes fully operational, it might be critical in driving the great interstellar ships. The materials needed for construction of energy beam facilities might most economically come from Mercury.

SETTLEMENT AT ALPHA CENTAURI

Interstellar travel is immensely difficult, and will be among the most technologically and economically difficult undertakings ever for our species. The best strategy cannot be determined yet, since it is dependent on the technology available, the cost, and the willingness to accept risk, including the probability of survival after arriving.

Alpha Centauri is the closest stellar system, and is likely to be the first reached by humans. The system consists of three stars. Two are sun-like, and are separated by about the distance from our Sun to Saturn. The third is a small red star, remote from the other two.

The First Starships

In one scenario, the first ships are either two-stage sailing ships or single-stage sailing ships with braking engines. A ship is driven by a laser energy beam located in solar orbit or on Mercury, aided by a focusing lens remotely located in solar space. Construction of a beam system is likely to take a few decades even after the design has been thoroughly developed.

The manned ships would carry adequate payload for survival and to begin a settlement. Robots and automated manufacturing machines will be some of the most important items carried.

A trip time of 100 years is often considered to be a practical upper limit for the crew, so the first trips might occur when the travel time can be reduced to no more than a century. Subsequent trips might have significantly shorter travel times.

A narrow beam would be blocked by a ship using it, so that only one ship could ride a narrow beam at a time. Thus a beam could be used only once per century, if that is the voyage time, unless a time-sharing strategy can be used. The strategy for the first voyage is then: 1) the first ship leaves when the first beam is ready, with each subsequent ship a few decades after its predecessor as new beams become available; 2) a fleet of 2 or 3 ships travel together when that many beams are ready, or a single wide beam is used; 3) a single ship makes the first voyage, with the second leaving when the beam is clear of the first; or 4) the ships time share a single beam, arriving a few years apart.

The great difficulty of reaching Alpha Centauri would be compounded if a return trip were to be made. Given this difficulty, the long voyage times, and the impetus to establish a settlement, the early voyages to Alpha Centauri are almost certain to be one-way only.

The ships will carry people, robots, supplies, equipment, and scientific instruments. The first people would establish a permanent settlement in the Alpha Centauri system. The primary tasks after arrival are naturally for survival and exploration. Beyond these, the priorities will likely be to build larger, more permanent living facilities, to build an economic infrastructure for their expanding population, and to possibly build facilities that would assist the arrival of subsequent ships.

Some of the ships would be used as solar sailing ships, providing transportation throughout the Alpha Centauri system. These ships could provide important transportation of materials to support the developing settlement. One application would be transportation of the products obtained from mining asteroids and comets. The key solar sailing regions around the three stars are shown in Figure 8.5.

The best strategy might be for the initial ship to be robotic. Its primary purposes would be exploration, habitat preparation, and possibly construction of a decelerator system to allow faster trip times when people make the voyage.

The case of using a deceleration beam for the first manned ship appears to have substantial advantages for safety, trip time, and possibly cost. This requires a robotic ship, or more than one, to construct or deploy a reflector that collects energy from the original beam and focuses it on approaching ships.

If the first arriving ship, either robotic or manned, is a sailing ship, its sail might be used as a decelerating reflector for following ships.

Buildup and Development

The rate of buildup depends on numerous factors, including the number of people and robots in the first ships and the frequency of additional arrivals. Using a deceleration beam and simple, single-stage ships should improve both of these factors.

The second generation ships may depart Earth only when the existing beams are sufficiently clear, so their arrival will be 50 to 100 years after the first ships. This makes it obvious that the first ship, and probably every other first generation ship, should be self sufficient for essentially permanent survival at the destination.

The second generation ships will carry people and generally the same type of items as the first. However, the mix of the cargo is hard to assess because of the uncertainty of the nature of the first generation of ships. Since the second generation may not arrive until a century after the first, they may not provide any critical benefit to the settlement if it is already successful. On the other hand, if the settlement is foundering, it may not last until the second generation arrives. Sophisticated logistics planning can help with this problem, but to a major extent, it may be part of the unavoidable risk in making the first voyages.

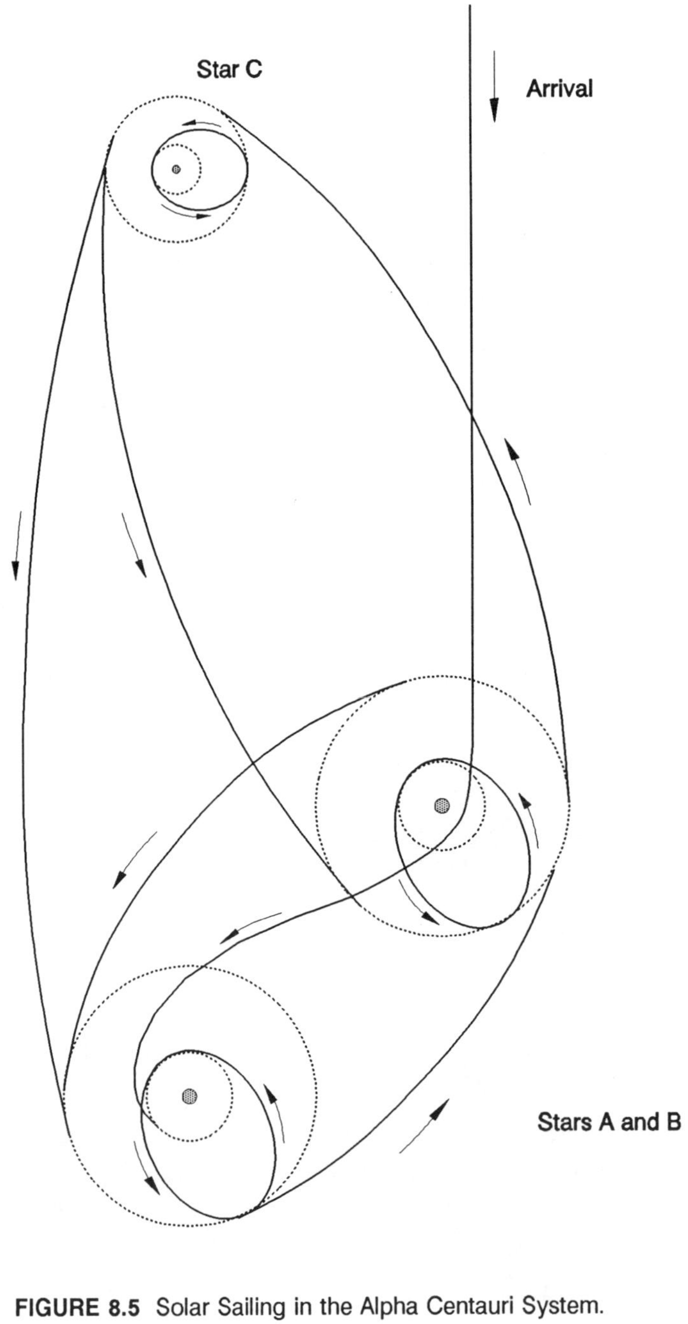

FIGURE 8.5 Solar Sailing in the Alpha Centauri System.

Having energy beams at Alpha Centauri to decelerate ships arriving from Earth could help reduce voyage times, perhaps to 30 years or less. From a transportation point of view, it would be very helpful for subsequent ships if a laser deceleration beam were quickly constructed. However, we cannot at present estimate the real difficulty of this task. If it takes a few decades to build a laser beam system back home, it could easily be a century or more until the new settlement could have the first beam system ready even if the beam had a high priority, which it may not. Subsequent beams might then become operational at intervals of a few decades.

A large reflector might be used instead of a laser. It would collect energy from the original beam and focus it on approaching ships. This reflector might be sent from Earth or built at the destination by robots or people.

If the decelerator is a laser rather than a reflector, there may be less operational constraints on the timing of ship arrivals, allowing ships to arrive more frequently. These third generation ships could be single stage ships relying on the decelerator system at the destination to reduce the trip time, minimize their cost, and to improve their reliability. The first of these ships seems unlikely to arrive sooner than about 150 years after the first ship. This simpler ship could not safely leave Earth only on the hope that the decelerator would be completed in time to stop it. If the decelerator is not available due to construction delay or failure, the approaching ship would become lost in interstellar space. As a result, the first of these ships would depart only after the decelerator has been completed and tested.

Like an accelerator beam, a narrow decelerator can have only one ship in the beam at a time. If the trip time is reduced to 30 years using accelerator/decelerator beams, them each beam pair could send a ship every 15 years. Further, if the laser beams are made sufficiently powerful that acceleration times are brief, then the beams can be time shared, with the result that ships could be arriving as often as every one to two years.

In addition to the work of further exploring and building up the settlement, the people would also work on expanding and maintaining the decelerator beam system. This will increase the flow of people and materials into the Alpha Centauri system, allowing a continuing expansion of activities. It could allow continuing reductions in travel time, so that even the more heavily loaded ships might arrive less than 30 years after leaving our solar system.

The survival needs of the settlement are of prime importance. The use of resources and the allocation of human and robotic labor must be directed to those needs. If significant resources and labor are needed to create a deceleration beam for future ships, that could be a poor choice if those ships will arrive so far in the future that they will not improve the probability of survival of the first settlers. In that case, it would be better for secondary ships to make longer trips while the first settlers concentrate on their survival.

It might be that a single ship could successfully establish a human settlement in Alpha Centauri, but for a long period of time, all interstellar effort is likely to be dedicated to getting additional ships there to aid the settlement. At some point, however, activities will be directed toward other stellar systems in our neighborhood. While the Alpha Centauri settlement transitions to a full, independent civilization, other starships will be departing Earth for other destinations and new civilizations. Eventually, these outlying civilizations will in their turn send out starships to destinations beyond the reach of Earth.

REFERENCES AND NOTES

1. This split mission concept, based on using sailing ships, to support a Mars expedition was invented by the author in 1980. It was publicly presented by Robert Staehle at the Case for Mars Conference, Boulder Colorado, 1981 April, and subsequently published: Staehle, R.L. (1982) An Expedition to Mars Employing Shuttle-Era Systems, Solar Sail and Aerocapture. *Journal of the British Interplanetary Society*, Vol. 35, pp. 327-335.

APPENDIX

A. SOLAR PRESSURE

Photons carry momentum, and when they are reflected, absorbed, or emitted, the interacting surface is subjected to a force caused by the change in the photon's momentum. A photon's momentum and energy are related by

$$p = h/\lambda = mc \tag{A.1}$$

$$E = mc^2 = pc \tag{A.2}$$

where p is the photon's momentum, h is Plank's constant, λ is the wavelength, m is the photon's relativistic mass, c is the speed of light, and E is the energy. The energy equation is valid for a single photon or a flux, thus the momentum of a light flux is given by

$$p = E/c \tag{A.3}$$

where E is now the energy of the flux. Using Newton's Second Law, the pressure experienced by a perfectly reflecting planar surface is then

$$P = \frac{2W \cdot \cos^2\alpha}{cR^2} = \frac{0.9126 \cdot \cos^2\alpha}{R^2} \times 10^{-5} \tag{A.4}$$

where P is in N/m^2, W is the intensity (1368 W/m^2 at 1 AU)[1], α is the angle between the surface normal and the Sun, and R is the solar distance in AU.[2]

The deviation of intensity from the inverse square of the distance may be ignored except when close to the Sun, when the effects of the solar sphere must be included.[3] The pressure exerted by the solar wind is 3 to 4 orders of magnitude less than photon pressure and is not considered further.

Real reflectors absorb some incident sunlight which is mostly in the visible wavelengths and radiate the absorbed energy away mostly in infrared wavelengths. The pressure on a perfectly absorbing surface perpendicular to the Sun, used as a reference pressure, is

$$P_a = \frac{4.563}{R^2} \quad \mu N/m^2 \tag{A.5}$$

Taking into account the reflected, absorbed, and radiated energy, the normalized force component normal to a planar surface is

$$\frac{F_N}{P_a A} = (1+rs)\cos^2\alpha + B_f r(1-s)\cos\alpha + \frac{B_f e_f - B_b e_b}{e_f + e_b}(1-r)\cos\alpha \tag{A.6}$$

The normalized force tangent to the surface is

$$\frac{F_T}{P_a A} = (1 - rs)\cos\alpha \cdot \sin\alpha \tag{A.7}$$

Then

$$F = (F_N^2 + F_T^2)^{1/2} \tag{A.8}$$

$$\tan\phi = \frac{F_T}{F_N} \tag{A.9}$$

B_f, B_b	-	non-Lambertian coefficients for front and back surfaces
A	-	surface element area
e_f, e_b	-	front and back surface IR emission coefficients
R	-	distance from Sun's center, in AU
r	-	reflectivity of front surface
s	-	specular reflection coefficient, fraction of reflected light
ϕ	-	angle of force vector from surface normal
α	-	angle of illumination from surface normal.

The reflectivity is the value for the wavelength of incident radiation, while the emissivities are for the wavelength of the emitted power, determined by the temperature of the sail.

The power emitted by both sides of the sail is accounted for in the preceding expressions. The power emitted from a surface of area A at temperature T is

$$Power = \sigma A T^4 e/a \tag{A.10}$$

where $\sigma = 5.67 \times 10^{-8}$ W m^{-2} K^{-4}, the Stefan-Boltzmann constant. The absorptance a is $1 - r$. The sail temperature is related to the solar intensity by

$$T^4 = \frac{a_f W}{(e_f + e_b)\sigma} \tag{A.11}$$

The effects from the cosmic background radiation at a temperature of 2.7 K is ignored under non-relativistic conditions. Under relativistic conditions, it should be included in evaluations, and can limit the speed of a sailing ship.[4]

The above equations include the factors thought to sufficiently determine the total force and moment on a sail. The results of numerical evaluations of the forces on sail surfaces are given in Appendix B based on these equations.

The coefficient of the first term in equation A.6 could more properly be written as $(1+xrs)$ where x is an efficiency factor to allow for dispersion of the specular reflection as a result of surface waviness. This waviness is due to small scale irregularities, creases, and distortions around seams. Measurements have indicated that this effect is less than 0.1%, justifying the simpler approach.

A comprehensive treatment of local factors determining the force on a sail leads to equations similar to the above, but more complex.[5] Many of the additional factors in these equations may be unmeasurable in a realistic manner. Any computation of forces and moments on a sail will include some uncertainties which can only be resolved by observing the performance of the ship during cruise conditions.

REFERENCES AND NOTES

1. Based on satellite (Solar Maximum Mission) measurements. The value used in 1976 was 1353 W/m^2.

2. Georgevic, R.M. (1973) The Solar Radiation Pressure Force and Torques Model. *The Journal of the Astronautical Sciences*, Vol. XX, No. 5, pp. 257-274, Mar-Apr.
3. McInnes, C.R. and Brown, J.C. (1989) *Solar Sail Dynamics with an Extended Source of Radiation Pressure*, International Astronautical Federation, IAF-89-350, October.
4. McInnes, C.R. and Brown, J.C. (1990) Terminal Velocity of a Laser-Driven Light Sail. *Journal of Spacecraft*, Vol. 27, No. 1, Jan-Feb.
5. Anselmo, V.J. (1977) *Derivation to Photon Force Equations for a Typical Solar Sail Material*, Jet Propulsion Laboratory, 900-847, Pasadena.

B. FORCES AND MOMENTS

The force and moment on a sail are obtained by performing a numerical integration of the solar pressure equations across the surface of the sail. This requires a sufficient knowledge of the interaction between the ship's structure and the sail that the sail shape can be accurately determined under load. In practice, this requires an iterative process since the distributed force on the sail is a function of the sail shape, and the shape is a function of the distributed force.

Integrating the F_N and F_T equations across the curved surface of the sail results in the determination of the total force and its angle ϕ with the centerline of the ship. This can be done by defining the sail surface to be made up of numerous flat facets that are oriented according to an assumed shape for the sail. The distributed force is then calculated, and the resultant deflection of structure and sail is then determined. If the resulting sail shape is not consistent with the assumed shape, an iterative process can be followed until convergence is obtained.

In the evaluation, the centerline of the ship is set at several angles α from the Sun line to determine the corresponding distributed forces and resulting moments. The angle between the force vector and the Sun line, shown in Figure B.1, is defined by

$$\theta = \alpha - \phi \tag{B.1}$$

Evaluations were made of a square-rigged sail and a heliogyro using the following parameter values:

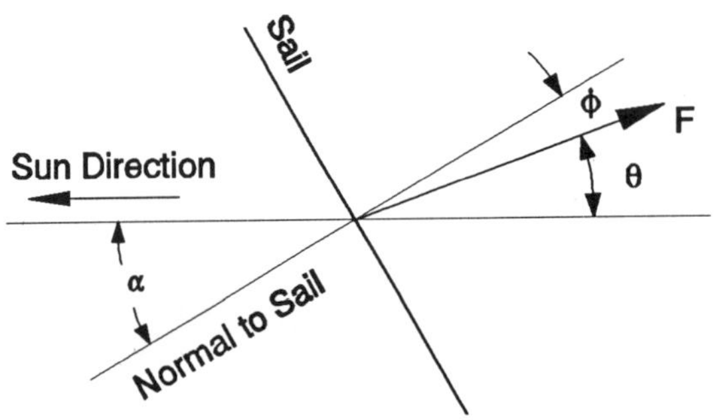

FIGURE B.1 Sail Directional Angles.

B_f	=	0.79	e_b	=	0.55
B_b	=	0.55	r	=	0.88
e_f	=	0.05	s	=	0.94

The resultant force functions and angles are plotted in Figures B.2 and B.3 for the square sail and heliogyro, respectively. These functions can be represented by

$$\frac{F}{F_0} = 0.349 + 0.662\cdot\cos 2\theta - 0.011\cdot\cos 4\theta \quad \text{(square sail)} \quad (B.2)$$

$$= 0.367 + 0.643\cdot\cos 2\theta - 0.010\cdot\cos 4\theta \quad \text{(heliogyro)} \quad (B.3)$$

where F_0 is the force for $\theta = \alpha = 0$.

These equations are needed for trajectory simulations of the ships, and are most conveniently expressed as functions of θ for those applications. The above expression for the heliogyro is valid when the sail blades are flat in the plane of rotation.

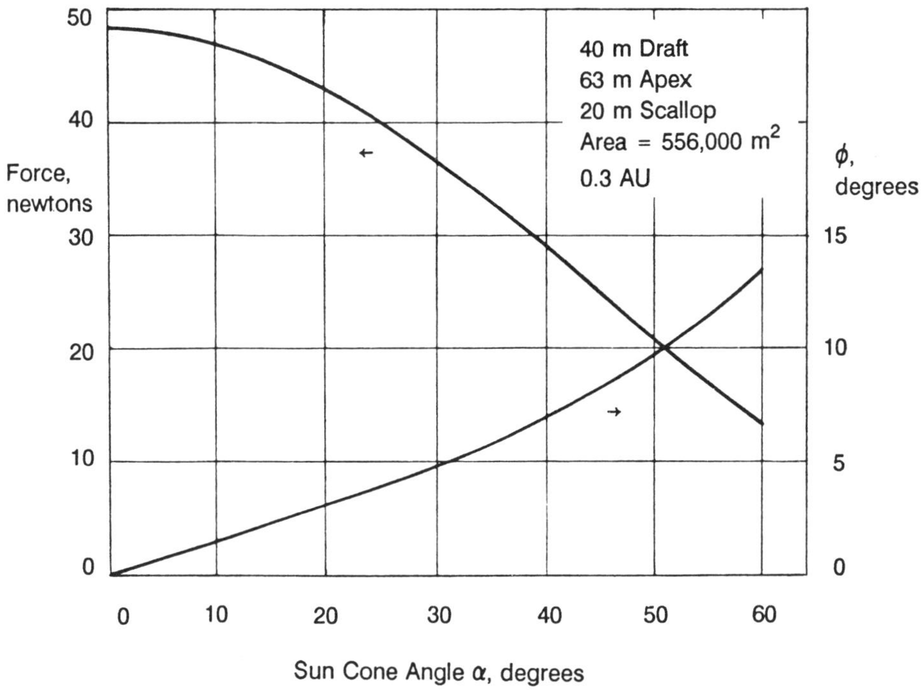

FIGURE B.2 Plot of Force Magnitude and Angle for Square Sail.

The moments on the 800 m square sail are shown in Figure B.4 as functions of the angle α for various heights of the central apex above the sail corners. Whenever a net moment exists, the ship will experience a rotation. Under steady cruise conditions, it is necessary to select an apex height that provides no net moment, or provide some other means of producing a balancing moment. It can be seen from the figure that pitch and yaw control of the ship can be obtained over a wide range of angles by changing the apex height, except when facing the Sun.

Corner vanes can be used to provide control moments for square sails, among others. The available normalized torques are shown in Figure B.5, using a coordinate system where the Z axis is normal to the sail.

The response of a square-rigged ship to the action of vanes with one degree of freedom in removing an attitude disturbance is shown in Figure B.6.

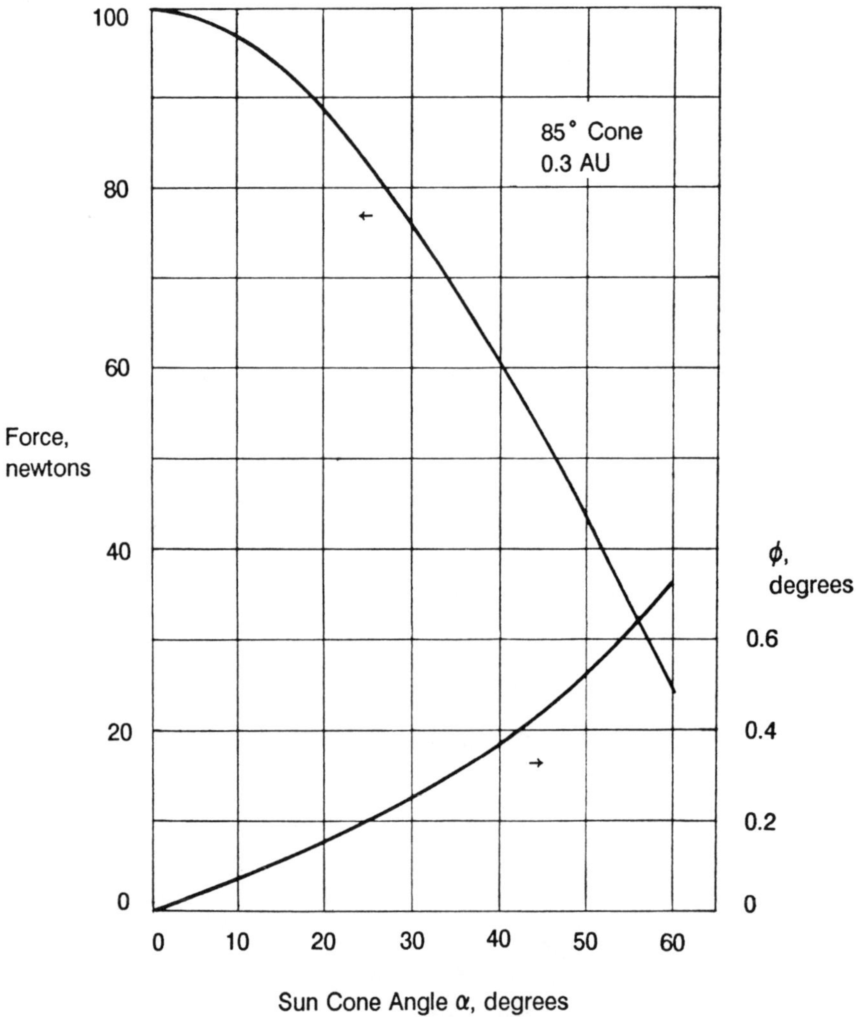

FIGURE B.3 Force Magnitude and Angle for Heliogyro.

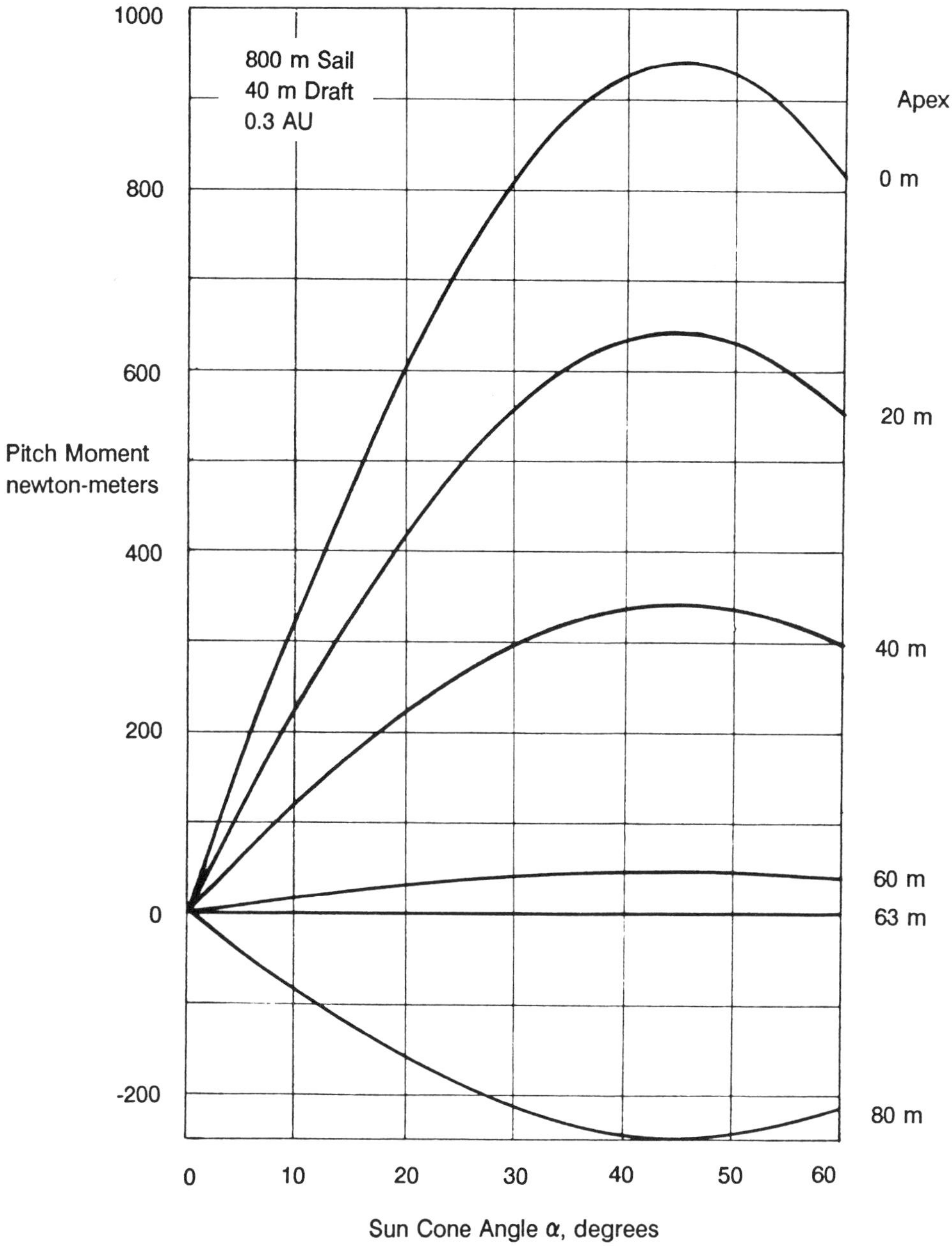

FIGURE B.4 Moments on a Square Sail.

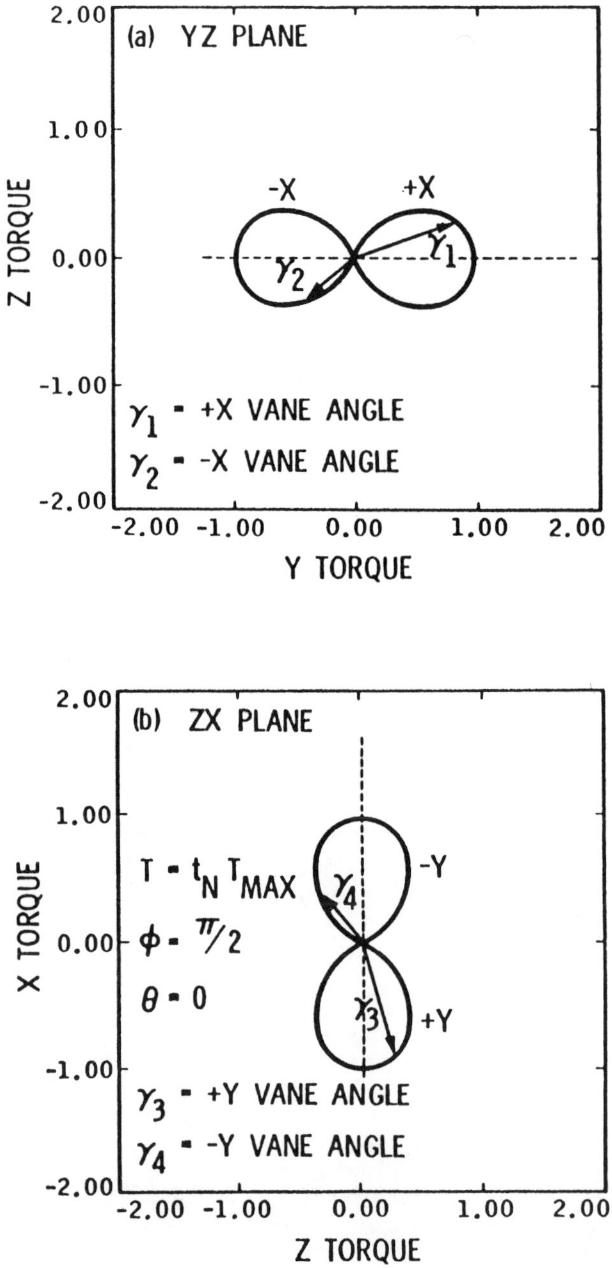

FIGURE B.5 Normalized Vane Torques for Square Sail. (JPL)

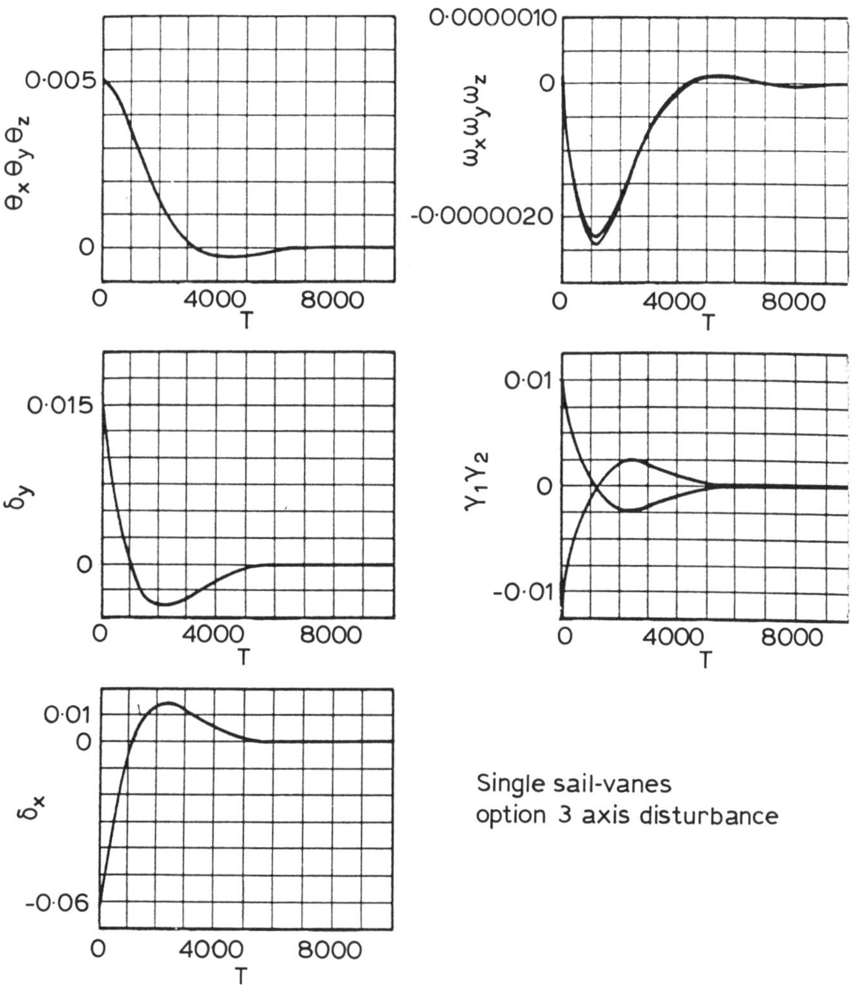

FIGURE B.6 Response of Square-Rigged Ship Using Single-Axis Vanes. (JPL)

C. TRAJECTORY ANALYSIS

The analysis of solar sailing trajectories depends on numerical integration of the equations of motion. Most interplanetary and planetary spiral trajectories are determined using a single, central body plus the spacecraft acting under a weak continual acceleration. Special situations could require including the effect of a third body, and including coast phases where there is no force on the sail to affect the trajectory. The key to a successful analysis is finding the minimum transfer time while meeting the necessary boundary conditions. Experience has shown that this is where the challenge lies.

INTERPLANETARY TRAJECTORIES

An analysis is done by application of Newton's Third Law of Motion, $F = ma$, with a model of the sail force, described in Appendix B, and a means of finding the best trajectory that meets the appropriate constraints.

A typical geometry of interplanetary trajectories is shown in Figure C.1. The direction of the force vector, given by θ, is the key parameter.

Use of simple accelerated motion in an inertial Cartesian system using a fourth-order Runga Kutta integrator is often the best approach to numerical integration of sailing trajectories.[1,2] The widely available personal computers are readily able to work these problems. The use of specialized coordinate frames and non-dimensional representations is not necessary.

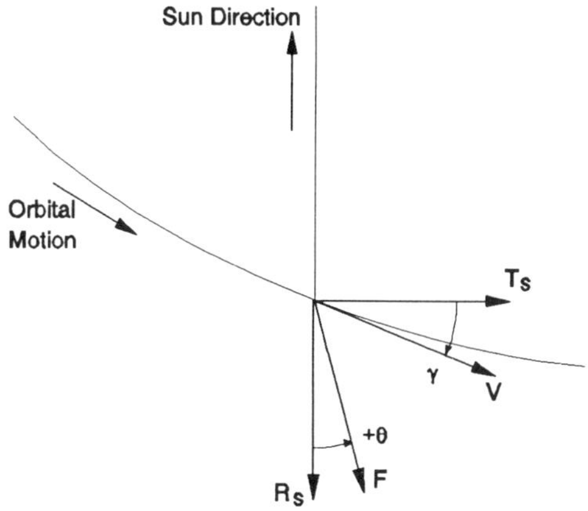

FIGURE C.1 Geometry of Velocity and Force Vectors.

Optimization by Method of Gradients

This approach is good for single and multiple parameters, and is relatively easy to set up. It has the advantage of using physically meaningful parameters. The analyst can provide good initial guesses based on experience and estimates. The results are generally acceptable for preliminary analysis and planning.

Previous work has been published for gradient[3,4] and conjugate gradient[5] techniques for Earth to Mars trajectories. A good introduction to several techniques with computer coded algorithms is also available.[6]

A simple, effective approach is to divide a trajectory into phases, each of which uses steering functions of the forms

$$\theta(t) = a + b \cdot t \tag{C.1}$$

$$\beta(t) = c + d \cdot t \tag{C.2}$$

where t is the time duration of the phase, θ is the pitch angle, and β is the clock angle around the Sun line. The optimization process determines a, b, c, and d.

These can be supplemented by special functions such as those for maximum instantaneous rate of energy gain or loss. The geometry of the velocity and force vectors for this steering mode is illustrated in Figure C.2.

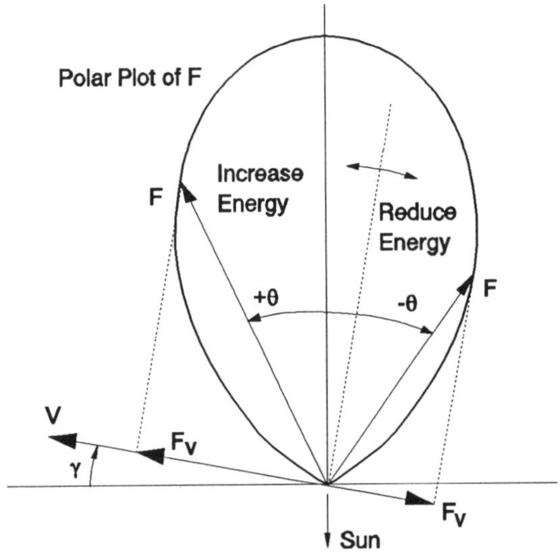

FIGURE C.2 Steering with Maximum Force Component On Velocity Vector.

A simple steering strategy can be used based on always maintaining the largest possible force component along the velocity vector. Table C.1 provides the relationship between the velocity angle, γ, and the force vector, θ, for an outward spiral.

An inward spiral case can make use of the same table, entering the table with a negated value of the velocity angle, $-\gamma$, and then negating the resulting angle, $-\theta$, of the force vector. As an example, an actual velocity angle of 10° requires a force angle of -35.5°.

A limit on the angle θ of 59° should be observed with this data. In the outward spiral case, the angle between $\mathbf{R_s}$ and **V** cannot exceed 149°. For an inward spiral, that same angle cannot be less than 31°. If these values are exceeded by the steering logic, the force magnitude must be nullified.

TABLE C.1 Steering Angles for Maximum Velocity Change.

Velocity Angle γ, *deg*	*Force Angle* θ, *deg*
90	0
80	2.9
70	5.9
60	9.0
50	12.2
40	15.5
30	19.1
20	22.6
10	26.7
0	30.8
-10	35.5
-20	40.4

The force magnitude is determined once the angle θ is determined. The equations for the force for the square sail and the heliogyro are given in B.2 and B.3, respectively. It should be remembered that F_0 varies by the inverse square of the distance from the Sun. For numerical simulations, it is convenient to use a vehicle mass of 1 kg and input the characteristic acceleration as the F_0 value.

Optimization by Calculus of Variations

The calculus of variations provides the most rigorous approach to optimization of sailing trajectories. It is a complex approach to implement, and it is often a difficult one to work with in achieving convergence on the boundary conditions. The quality of the results is usually substantially better than those obtained from gradient techniques. A full treatment of the calculus of variations approach has been done by Carl Sauer using realistic sail models and actual ephemerides.[7]

PLANETARY SPIRAL ORBITS

Planetary spirals are more difficult to determine than interplanetary trajectories because of the more complex geometry. The relative geometry of the solar, force, and velocity vectors is shown in Figure C.3, given with

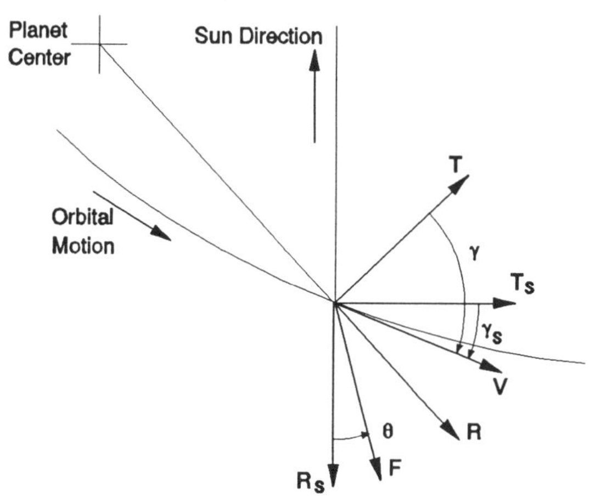

FIGURE C.3 Geometry of Velocity and Force Vectors in a Planetocentric Orbit.

respect to the usual radial-tangential-normal (RTN) coordinate frame.

The force vector must be nullified when in the shadow region. The mean solar rate (mean Earth orbital rate) is 0.9856° per day. Other factors that can be included for increased accuracy include albedo effects, non-spherical gravitational effects, and perturbations by other bodies.

The relationship of the force vector and velocity vector given in Table C.2 may be used in a simple, effective steering strategy based on maintaining the maximum energy gain or loss at all times.

It is not necessary to integrate along an entire escape spiral. A single orbit can be numerically integrated, and the change in orbital elements during that orbit can be projected ahead for some reasonable number of orbits. The projected orbit is also numerically integrated to determine the current rate of change in orbital elements per orbit, enabling another projection to be made. This process can be continued, with progressively smaller steps between orbits as the orbits increase in size. Eventually, each consecutive orbit must be numerically integrated until escape conditions are achieved.

Sailing spirals have significant eccentricities most of the time, which may complicate the optimization process. An unconstrained optimization procedure can easily select orbits that intersect the planet.

Transition Region and Optimization

The transition region includes the sphere of influence and the associated zones of perturbations. Gravitational sources are the planet, the Sun, and possibly satellites of the planet.

Escape trajectories through the transition region are influenced by whether the interplanetary trajectory is inbound or outbound. The optimization of the path through the outer part of the planetary environment must take the subsequent interplanetary trajectory into account.

The spirals can have high eccentricities, which can be used to advantage to the extent that the orbital orientation can be controlled. Of course, this adds more complexity to an already difficult optimization problem.

Outbound trajectories have the advantage of being able to run directly away from the Sun as the total energy becomes positive. This makes the last few orbits very effective for energy gain, which minimizes the spiral time and provides a substantial velocity as the ship crosses the sphere of influence.

Inbound trajectories have the problem of wanting to depart toward the

Sun, which is the source of the radially outward force. The ship needs to gain energy with respect to the planet while it needs to lose energy with respect to the Sun. The ship must coast through the region where adding final escape energy is most effective, where it is more or less approaching the Sun. It can begin the final energy gain when it passes between the planet and the Sun, moving retrograde with respect to the Sun. However, the sail must be at an angle to the Sun, so the energy gain is not as rapid as for an outbound trajectory. The result is that the spiral time is longer and the velocity crossing the sphere of influence is not as great as in the outbound case.

Capture maneuvers can be done with two basic approaches. In the first, the ship is allowed to cross the sphere of influence with a significant velocity, from the anti-Sun direction. On the leading side of the planet, the sail is turned more or less directly toward the Sun providing an effective braking maneuver which often results in a highly eccentric orbit with periapse on the leading side of the planet.

The second approach is essentially a true rendezvous, where the ship has a near-zero velocity at the sphere of influence. This rendezvous point can be placed almost anywhere around the planet, providing flexibility in selecting the orientation of the capture orbit. This comes at the expense of a longer interplanetary transfer and a longer time spent in entering the capture orbit.

REFERENCES AND NOTES

1. Sauer, C.G. (1976) *Optimum Solar Sail Interplanetary Trajectories*. American Institute of Aeronautics and Astronautics, 76-792, New York.
2. Press, W.H., Flannery, B.P., Teukolsky, S.A., Vetterling, W.T. (1986) *Numerical Recipes*. Cambridge University Press, Cambridge, pp. 274-334.
3. Kelley, H.C. (1960) Gradient Theory of Optimal Flight Paths. *Journal of the American Rocket Society*, Vol. 30, Oct., pp. 947-954.
4. Kelley, H.C. (1962) Methods of Gradients. *Optimization Techniques*, ed. Lietman, Academic Press, New York, pp. 205-254.
5. Jayaraman, T.S. (1980) Time-Optimal Orbit Transfer Trajectory for Solar Sail Spacecraft. *Journal of Guidance and Control*, Vol. 3, No. 6, pp. 536-542.
6. Press, W.H., Flannery, B.P., Teukolsky, S.A., Vetterling, W.T. (1986).
7. Sauer, C.G. (1976).

D. HELIOGYRO DYNAMICS

The dynamics of the heliogyro are the greatest concern in its design and operation. The impact of dynamics on the design is 20% of the sail module mass. Components required by rigid body dynamics are pitch actuators and an autopilot. Components required by flexible body dynamics are dampers and flap hinge braces. Components partly designed by dynamics are the edge tendons, the deployment booms and yokes, and the center body structure.

The principal dynamics concerns are:

- maneuver rates
- suppression of nutation
- damping of all blade modes
- dynamic loads during maneuvers
- effects of dimensional tolerances
- blade flutter
- deployment dynamics
- damage tolerance
- multiple interaction of individual concerns.

The data and figures contained in this section are from R.E. MacNeal.[1]

RIGID BODY DYNAMICS

In one class of dynamics problems, the heliogyro is treated as a rigid body. This assumption allows a simplification of the analysis while leaving adequate accuracy. Maneuver requirements, blade pitch, and nutation are the issues addressed under rigid body dynamics.

Maneuver Requirements

Deployment requirements are to unroll the blades and add angular momentum to the spacecraft as the blades unroll, a process taking approximately 12 days.

Cruise requirements are to change the vehicle attitude to maintain the correct orientation with respect to the Sun, which is done at rates less than 8° per day.

Cranking orbit requirements, part of the Halley Rendezvous mission, are to maintain Sun pointing (precess at 8° per day) and to change the thrust direction twice per orbit, precessing 70° in about one day.

Rendezvous requirements are to control the thrust vector magnitude and direction (change 10% in 4 hours) during the final approach.

Planetary escape requirements are to change the thrust magnitude and direction (10-100%) twice per orbit.

Blade Pitch Modes

All control techniques for the heliogyro rely on changing the pitch angle of the blades. These changes are done in three modes.

Collective pitch, shown in Figure D.1, is the mode where all blades have the same blade pitch angle simultaneously. The effect on the vehicle is to alter the spin rate through the resultant torque. The total force acting normal to the disk is reduced while the collective pitch exists.

The half-period (half-p) pitch mode, Figure D.1, has blades pitching in opposite orientations on alternate revolutions. In this mode, a standing blade pitch pattern exists at a selected clock position in the plane of rotation. As a given blade passes that position, it will have a selected pitch angle, and on the subsequent pass it will have an equal but opposite pitch angle. The result is that the spin rate is not affected, but the solar pressure is reduced on the disk where the standing blade pattern exists. This creates a torque on the disk, causing the vehicle to precess. This mode was preferred for the Halley Rendezvous mission.

The cyclic (full period) pitch mode, shown in Figure D.2, has the blades

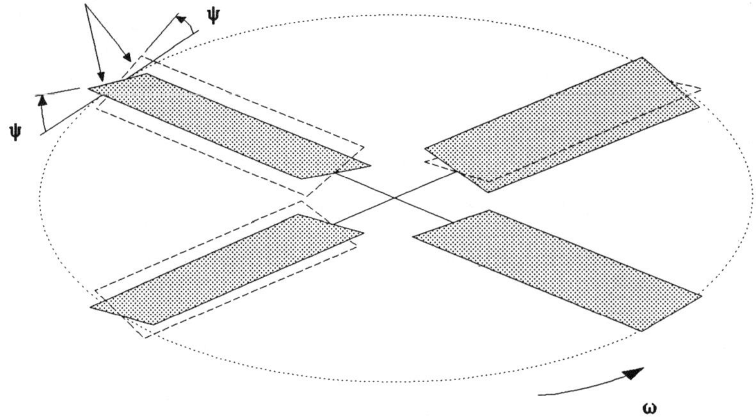

a) Cyclic Mode (Half Period)

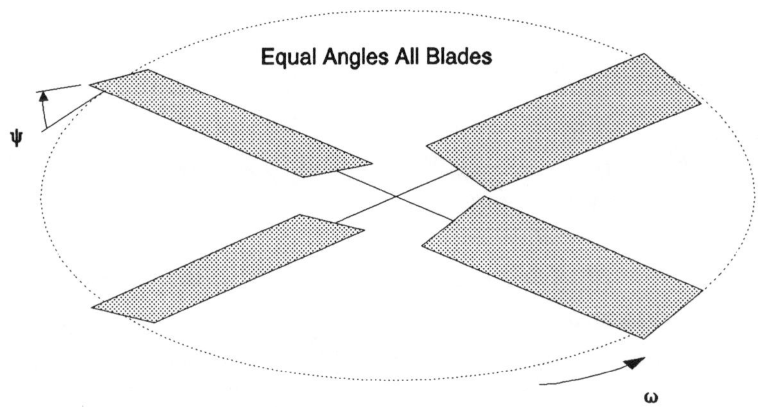

b) Collective Mode

FIGURE D.1 Half-Period and Collective Pitch Modes.

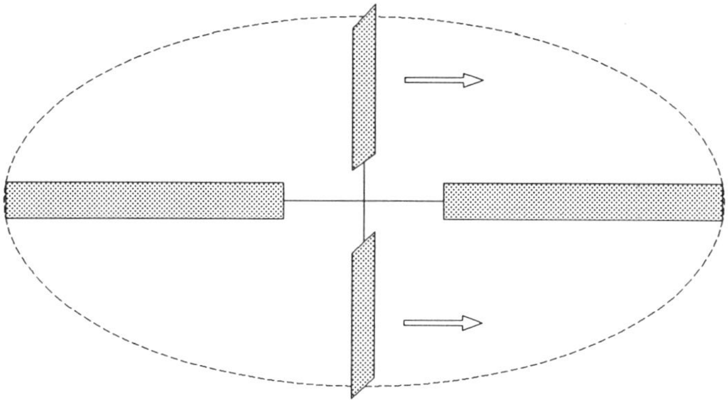

FIGURE D.2 Cyclic (Full-Period) Pitch Mode.

pitching in opposite orientations within a single revolution. A lateral thrust component is generated without any torque to alter the spin rate or cause precession. This can be useful in planetary spiral orbits and in rendezvous maneuvers, but it was found to not be necessary for the Halley mission.

In this mode, the total thrust can be reduced as much as 20%, or the thrust vector can be turned by 6°. Both of these changes can be accomplished in about 10 minutes.

A summation of blade pitch angles needed to meet the Halley Rendezvous mission requirements is given in Table D.1 for the different modes.

Nutation

The nutation motion is a wobble in the rotor disk at 2 cycles per revolution. The best method of control is to use the half-period pitch mode with appropriate phasing between the blades. Instead of the standing wave pattern used for vehicle precession, this uses a forward traveling wave pattern. The nutation can be a continuous problem, so the autopilot must be able to integrate this pattern of blade control with any other control requirements.

TABLE D.1 Pitch Angles Necessary to Meet Halley Rendezvous Maneuver Requirements.

Maneuver	*Half-Period* $a_{1/2}$	*Collective* Ψ_c	*Cyclic* a_1	b_1
Spin-Up		35°		
Cruise	10°			
Slow Precession[a]	10°	<1°		
Rapid Precession[b]	26°	4°		
Thrust Modulation[c]	±21°		12°	12°

[a]Precess 360° in 45 days
[b]Precess 70° in 24 hours
[c]Reduce thrust up to 20% and change direction by 6° in 10 minutes

FLEXIBLE BODY DYNAMICS

The blades do not act like rigid structures, but instead undergo complex deflections that depend on the rotation rate, disturbing forces, and solar pressure.

The identified problems can be categorized as static distortions, vibrations, and instabilities. The detection of these problems, and the verification of solutions, depends mainly on analysis and computer simulation. Some model testing could be done, but it is not possible to match the parameters of the actual sail, leaving a need for substantial further analysis to understand the correlation of the model tests with actual dynamic behavior of the sail in space.

Pitch Dynamics

The only torsion available to twist the blades comes from the dynamics of rotation. The angle of twist that can be achieved at the tip is less than the angle at which the root is set in static conditions, as indicated in Figure D.3. The static case corresponds to the fixed collective pitch mode.

The roots of the blades can be given sinusoidal excitations at less than or greater than the rotational speed of the disk in the non-static case. The ratio to the two rates determines the behavior of the blade tip relative to the root. The ratio of tip angle to root angle is shown in Figure D.4 under steady state conditions.

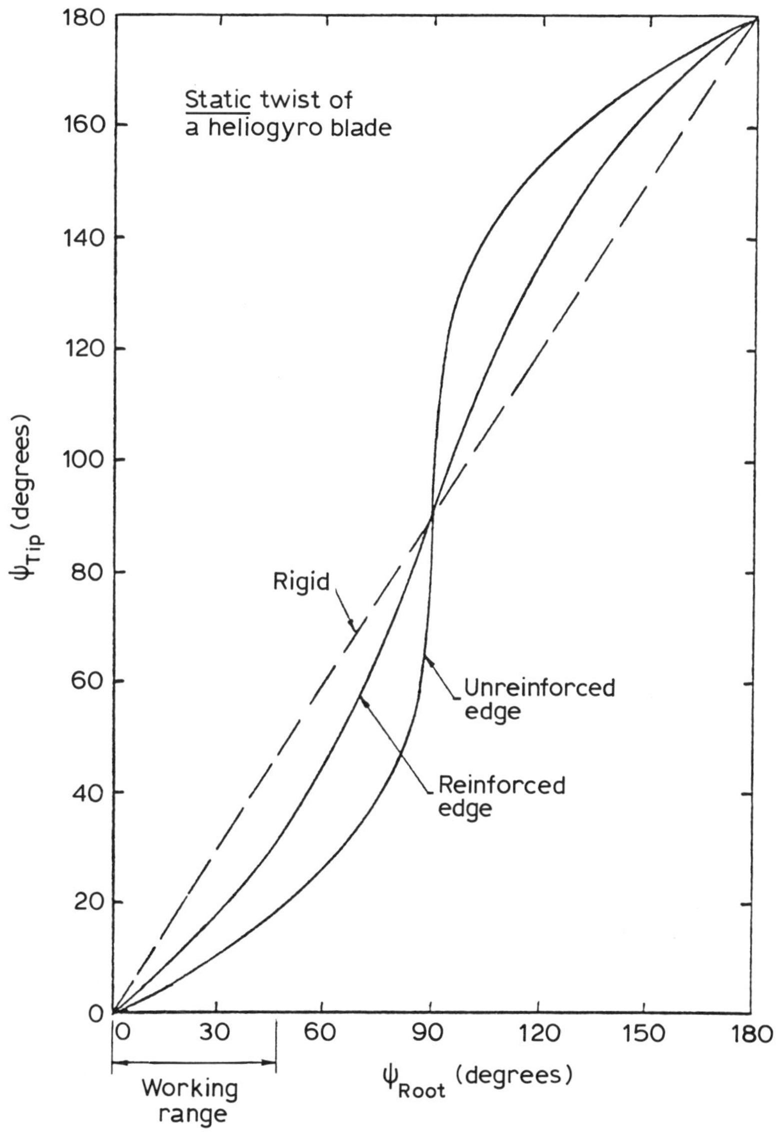

FIGURE D.3 Static Twist of the Heliogyro Blade. (JPL)

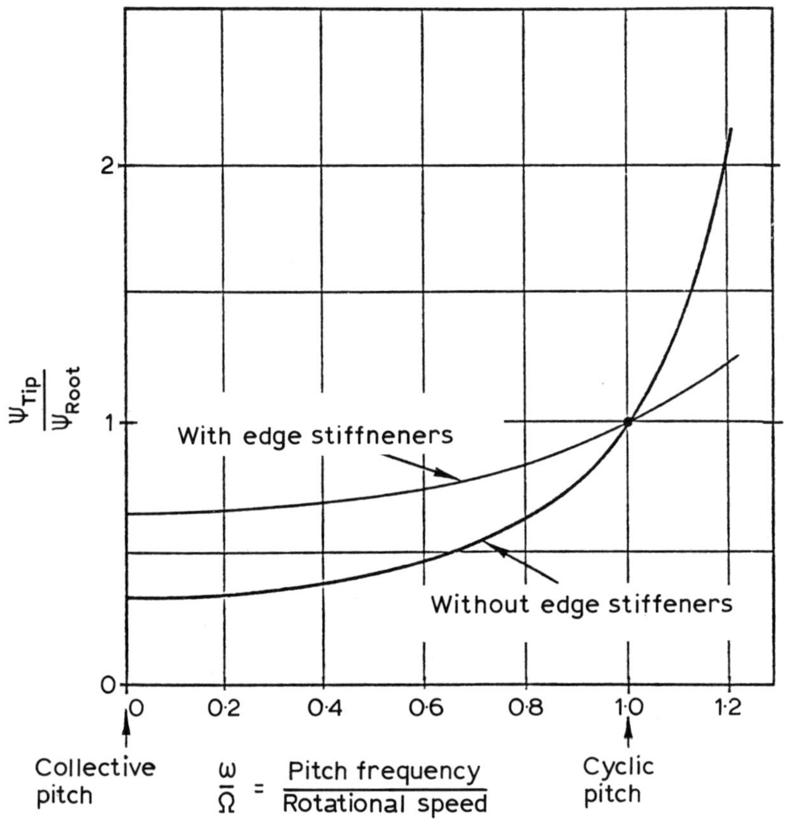

FIGURE D.4 Steady State Blade Tip Pitch. Response to sinusoidal excitation at the blade root, assuming small motions. (JPL)

There is a time lag between the commanding of blade twist and the response of the blade, which varies along the length of the blade. The behavior of the tip and root cords are shown in Figure D.5. Approximately half of a revolution is required for the blade root to reach its nominal twist angle. The tip of the blade will undergo oscillations that require 5 or more revolutions before reaching its final twist angle.

The dynamic pitch at the tip in response to a ramped twisting of the root to a fixed angle is shown in Figure D.6 for the case of collective pitch.

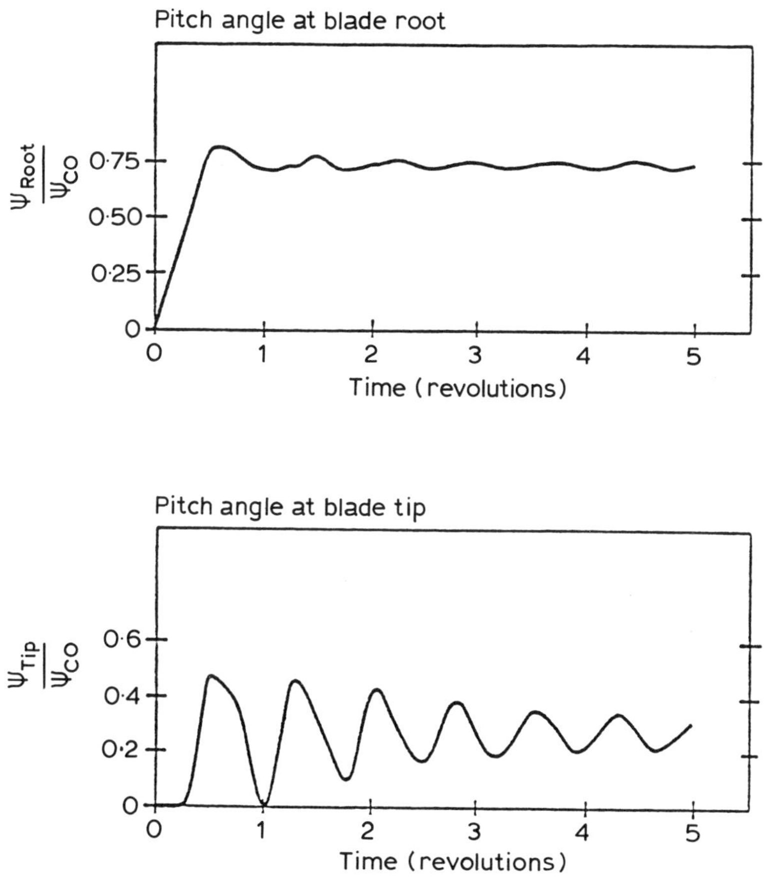

FIGURE D.5 Transient Torsional Response of Blade to Ramp Input. (JPL)

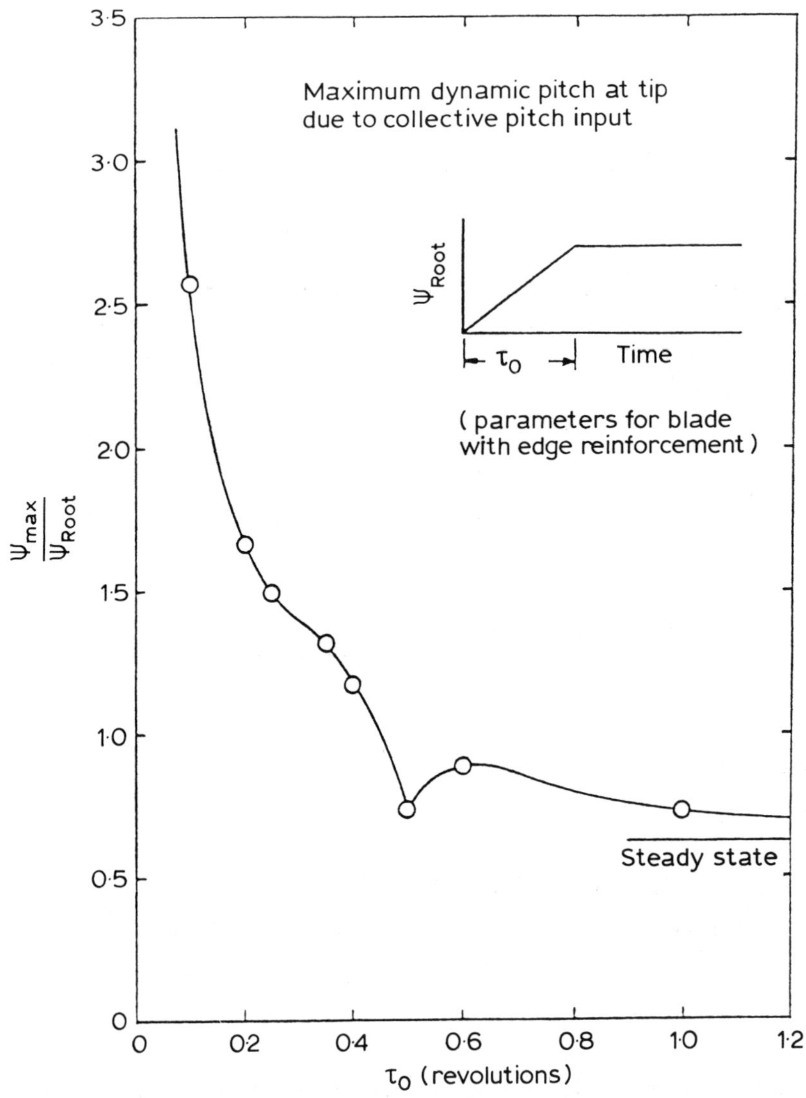

FIGURE D.6 Blade Tip Response to Ramped Collective Pitch. (JPL)

Vibration Modes

The blades can experience three types of vibrations. Lag bending is the blade tip lagging behind or surging ahead of its nominal position radially out from the root position. Flap bending is the motion of the blade tip above and below the plane of rotation. Torsional bending is the twisting of the blade. The frequency spectra of these modes are shown in Figure D.7.

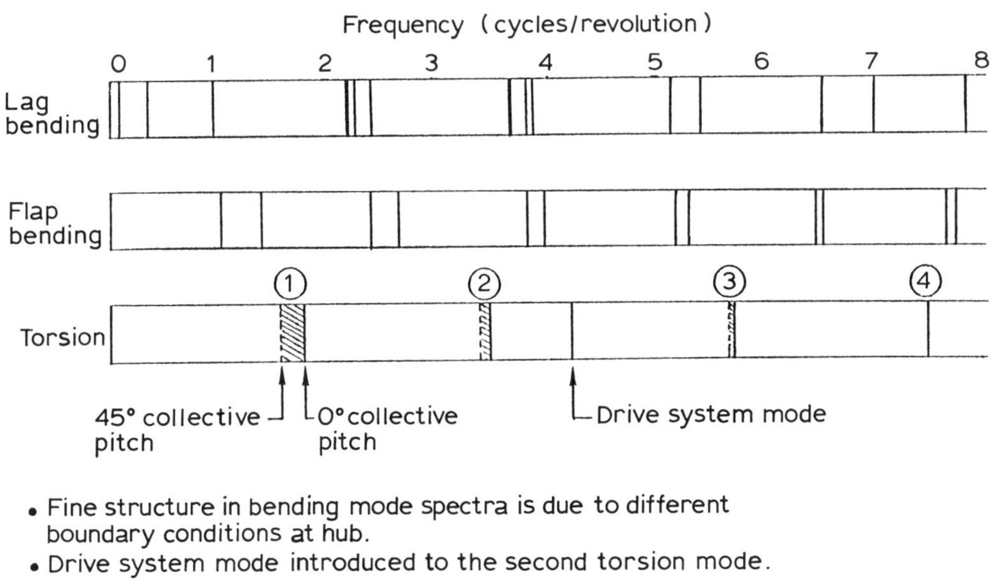

FIGURE D.7 Heliogyro Vibration Mode Frequency Spectrum. (JPL)

Damping

Damping is required to limit vibratory responses to excitations (transient, periodic, or random) and to suppress instabilities. The blades themselves have no inherent damping against torsion and flapping motions. The inherent damping against lag motion is very low. As a result, damping capabilities must be added. This is best done with elastomeric dampers placed in the flap hinge brace. Damping parameters are shown in Table D.2.

TABLE D.2 Damping Levels of Hub Fixed Modes.

Mode	*Frequency cycles/rev*	$g = Q^{-1}$	*Time to Half Amplitude, hrs*
1st Torsion	1.77	0.0025	3.2
1st Flap	1.0034	0.00081	17.5
2nd Flap	2.419	0.00071	8.3
1st Lag	0.0863	0.1487	1.1
2nd Lag	2.205	0.00116	5.6

Dynamic Loads

Severe maneuver requirements mean that the center body and the disk of rotation must be reoriented at high rates. The load effects from starting and stopping these maneuvers were investigated in the context of the Halley Rendezvous mission requirements. The responses of the blades were found to be small relative to the displacements that could cause adverse blade behavior, such as an edge tendon becoming slack. The dynamic loads spectra are shown in Figure D.8, and the blade response to dynamic loads is shown in Figure D.9. The figures shown that loads diminish at higher frequencies.

Dynamic Instabilities

An investigation of thermal effects on the blades showed that thermo-elastic instabilities will not occur because of small temperature differences between leading and trailing edges, short thermal lag times, and small coefficients of thermal expansion.

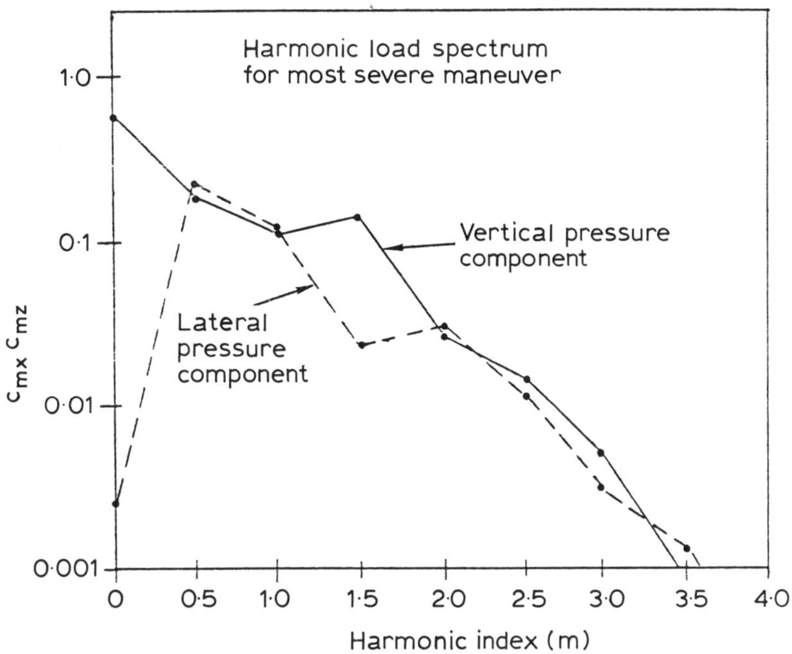

$$p_x = p_o \sum_m c_{mx} \sin(m\Omega t + \phi_{mx})$$

$$p_z = p_o \sum_m c_{mz} \sin(m\Omega t + \phi_{mz})$$

FIGURE D.8 Dynamic Loads Spectrum. (JPL)

The use of two (or more) banks of blades can require the center body to be long relative to its diameter. This is the case with the Halley Rendezvous design, which does not become stable in spin until the blades are deployed a minimum of 150 m. Prior to that extent of blade deployment, the dampers might need to be locked to avoid an upset in the rotation of the vehicle.

Flutter is the primary dynamic concern. It is caused by mechanical coupling while flap and lag motions exist, illustrated in Figure D.10. The mechanical coupling is associated with hub tilt (causes whirl-mode flutter) and lack of blade straightness due to light pressure deflections and differential strain between leading and trailing edges.

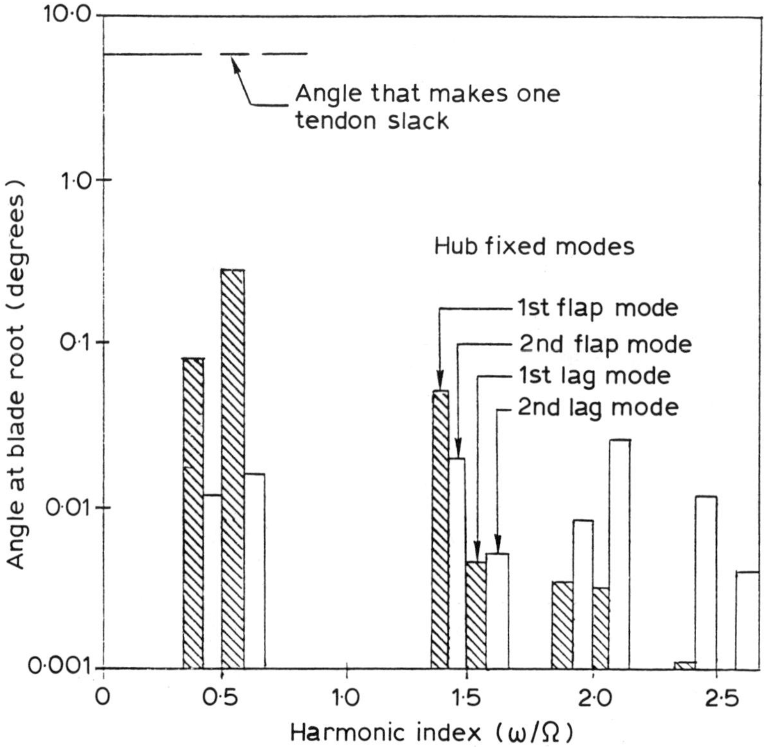

FIGURE D.9 Blade Response to Dynamic Loads. (JPL)

Several approaches are available to control blade flutter. The hub tilt problem can be controlled by an 80% decoupling of blade pitch from the hub tilt using an electronic swash plate. Blade stiffness can be increased by increasing the centripetal effect with a faster rotation. Damping is required for each mode of blade motion, and the mechanical design must keep a separation between the frequencies of the lower bending and torsion modes. Control actions must avoid adverse amplitudes and frequencies of blade pitch.

A structure-autopilot instability is a definite hazard because of the flexibility and response times of the structural system. It can be prevented by well-designed mechanical damping for all vibration modes and by adequate gain margin for all vibration modes. This can be achieved by a steep gain roll-off plus notch filters for the lower modes.

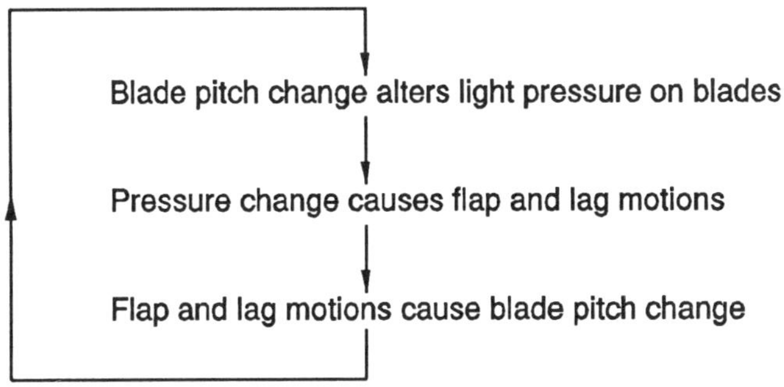

FIGURE D.10 Mechanisms of Blade Flutter.

Dynamic problems during deployment are primarily associated with lag and flapping. The Coriolis effect causes blade lag, which limits the initial deployment rate to 200 m in 20 minutes. Flutter damping is reduced by a factor of two because the flap hinge brace is not erected until the blade is fully deployed. However, there should be no major problems since no maneuver requirements are placed on the vehicle during deployment except to maintain pointing at the Sun.

The possibility of multiple interactions of individual dynamic concerns was identified as a potential problem. However, curtailment of the solar sail effort at JPL prevented an analysis from being done.

REFERENCES AND NOTES

1. MacNeal, R.E. (1977) Heliogyro Dynamics. *Solar Sail Technology Readiness Report*, Jet Propulsion Laboratory, Report 720-1, Pasadena.

INDEX